Electronics
DeMYSTiFieD®

DeMYSTiFieD® Series

Accounting Demystified
Advanced Calculus Demystified
Advanced Physics Demystified
Advanced Statistics Demystified
Algebra Demystified
Alternative Energy Demystified
Anatomy Demystified
asp.net 2.0 Demystified
Astronomy Demystified
Audio Demystified
Biology Demystified
Biotechnology Demystified
Business Calculus Demystified
Business Math Demystified
Business Statistics Demystified
C++ Demystified
Calculus Demystified
Chemistry Demystified
Circuit Analysis Demystified
College Algebra Demystified
Corporate Finance Demystified
Databases Demystified
Data Structures Demystified
Differential Equations Demystified
Digital Electronics Demystified
Earth Science Demystified
Electricity Demystified
Electronics Demystified
Engineering Statistics Demystified
Environmental Science Demystified
Everyday Math Demystified
Fertility Demystified
Financial Planning Demystified
Forensics Demystified
French Demystified
Genetics Demystified
Geometry Demystified
German Demystified
Home Networking Demystified
Investing Demystified
Italian Demystified
Java Demystified
JavaScript Demystified
Lean Six Sigma Demystified

Linear Algebra Demystified
Macroeconomics Demystified
Management Accounting Demystified
Math Proofs Demystified
Math Word Problems Demystified
MATLAB® Demystified
Medical Billing and Coding Demystified
Medical Terminology Demystified
Meteorology Demystified
Microbiology Demystified
Microeconomics Demystified
Nanotechnology Demystified
Nurse Management Demystified
OOP Demystified
Options Demystified
Organic Chemistry Demystified
Personal Computing Demystified
Pharmacology Demystified
Physics Demystified
Physiology Demystified
Pre-Algebra Demystified
Precalculus Demystified
Probability Demystified
Project Management Demystified
Psychology Demystified
Quality Management Demystified
Quantum Mechanics Demystified
Real Estate Math Demystified
Relativity Demystified
Robotics Demystified
Sales Management Demystified
Signals and Systems Demystified
Six Sigma Demystified
Spanish Demystified
sql Demystified
Statics and Dynamics Demystified
Statistics Demystified
Technical Analysis Demystified
Technical Math Demystified
Trigonometry Demystified
uml Demystified
Visual Basic 2005 Demystified
Visual C# 2005 Demystified
XML Demystified

Electronics
DeMYSTiFieD®

Stan Gibilisco

Second Edition

New York   Chicago   San Francisco   Lisbon   London   Madrid   Mexico City
Milan   New Delhi   San Juan   Seoul   Singapore   Sydney   Toronto

The McGraw·Hill Companies

Cataloging-in-Publication Data is on file with the Library of Congress

McGraw-Hill books are available at special quantity discounts to use as premiums and sales promotions, or for use in corporate training programs. To contact a representative please e-mail us at bulksales@mcgraw-hill.com.

Electronics DeMYSTiFieD®, Second Edition

1 2 3 4 5 6 7 8 9 0   DOC/DOC   1 9 8 7 6 5 4 3 2 1

ISBN 978-0-07-176807-8
MHID 0-07-176807-6

Sponsoring Editor	**Copy Editor**
Judy Bass	Nancy Dimitry,
	D&P Editorial Services, LLC
Editing Supervisor	
David E. Fogarty	**Proofreaders**
	Don Pomeranz, Don Dimitry
Production Supervisor	D&P Editorial Services, LLC
Pamela A. Pelton	
	Art Director, Cover
Project Manager	Jeff Weeks
Joanna Pomeranz,	
D&P Editorial Services, LLC	**Cover Illustration**
	Lance Lekander
Composition	
D&P Editorial Services, LLC	

To Tim, Tony, and Samuel
from Uncle Stan

About the Author

Stan Gibilisco, an electronics engineer and mathematician, has authored multiple titles for the McGraw-Hill *Demystified* and *Know-It-All* series, along with numerous other technical books and dozens of magazine articles. His work has been published in several languages.

Contents

Introduction

This book can help you learn basic electronics without taking a formal course. It can also serve as a supplemental text in a classroom, tutored, or home-schooling environment. None of the mathematics goes beyond the high-school level. If you need a refresher, you can select from several *Demystified* books dedicated to mathematics topics. If you want to build yourself a substantial mathematics foundation before you start this course, I recommend that you read and study *Algebra Know-It-All* and *Pre-Calculus Know-It-All*.

How to Use This Book

This book contains abundant multiple-choice questions written in standardized test format. You'll find an "open-book" quiz at the end of every chapter. You may (and should) refer to the chapter texts when taking these quizzes. Write down your answers, and then give your list of answers to a friend. Have your friend tell you your score, but not which questions you missed. The correct answer choices are listed in the back of the book. Stay with a chapter until you get most of the quiz answers correct.

Three major sections constitute this course. Each section ends with a multiple-choice test. Take these tests when you're done with the respective sections and have taken all the chapter quizzes. The section tests are somewhat easier than the chapter-ending quizzes. The course concludes with a final exam. Take it after you've finished all the sections, all the section tests, and all the chapter quizzes. You'll find the correct answer choices for the section tests and the final exam listed in the back of the book.

With the section tests and the final exam, as with the quizzes, have a friend divulge your score without letting you know which questions you missed. That way, you won't subconsciously memorize the answers. You might want to take each test, and the final exam, two or three times. When you get a score that makes you happy, you can (and should) check to see where your strengths and weaknesses lie.

I've posted explanations for the chapter-quiz answers (but not for the section-test or final-exam answers) on the Internet. As we all know, Internet particulars change; but if you conduct a phrase search on "Stan Gibilisco," you'll get my Web site as one of the first hits. You'll find a link to the explanations on that site. As of this writing, it's **www.sciencewriter.net**.

Strive to complete one chapter of this book every 10 days or two weeks. Don't rush, but don't go too slowly either. Proceed at a steady pace and keep it up. That way, you'll complete the course in a few months. (As much as we all wish otherwise, nothing can substitute for "good study habits.") When you're done with the course, you can use this book as a permanent reference.

I welcome your ideas and suggestions for future editions.

Stan Gibilisco

Part I

Fundamental Concepts

chapter 1

Direct Current

Engineers use the term *direct current* (DC) to describe a flow of *charge carriers* in an unchanging direction. A charge carrier is an object or particle with a specific, known *electrical charge*. The most common charge carrier is the *electron*, a subatomic particle. Electrons move from atom to atom in substances called *electrical conductors*. In the known universe, every electron carries the same charge quantity as every other electron: one negative *elementary charge unit* (−1 ECU).

CHAPTER OBJECTIVES

In this chapter you will

- Learn how direct current and voltage relate to resistance and conductance.
- Define and express Ohm's Law.
- Perform simple calculations using Ohm's Law.
- Learn about series and parallel connections.
- Discover how resistor values combine when you connect them together.
- Review the basic principles of DC magnetism.

The Nature of DC

Figure 1-1 illustrates four simple graphs of electrical current versus time. The graphs at A, B and C depict DC because the current always flows in the same direction, even though the amplitude (intensity) does not necessarily remain constant. The rendition at D does not illustrate DC because the current does not always flow in the same direction.

What is Current?

Electrical current quantifies the rate at which charge carriers flow into, out of, or past a specific point in a circuit. The standard unit of current is the *ampere*, which engineers abbreviate as A. When you observe a current of 1 ampere (1 A) at some point, you witness the movement of 1 *coulomb* (6.24×10^{18}) of charge carriers past that point every second.

You'll often want to specify current in smaller units than the ampere. You can use *milliamperes*, abbreviated mA, where

$$1 \text{ mA} = 0.001 \text{ A}$$

You'll also sometimes read or hear of *microamperes* (μA), where

$$1 \text{ μA} = 10^{-6} \text{ A}$$
$$= 0.001 \text{ mA}$$

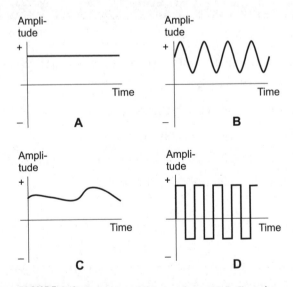

FIGURE 1-1 · Examples of DC waveforms (A, B, C), and a non-DC waveform (D).

Once in a while, you'll encounter *nanoamperes* (nA), where

$$1 \text{ nA} = 0.001 \text{ } \mu\text{A}$$
$$= 10^{-9} \text{ A}$$

A current of a few milliamperes through your body will give you a significant electrical shock; 50 mA will jolt you severely; 100 mA can kill you if it follows a path through your heart.

In some circuits, extremely large currents flow. Imagine a thick, solid copper bar placed directly across the output terminals of a large electric generator. The generator can drive huge numbers of electrons through the bar every second, representing a current that can exceed 100 A. In other circuits, the flow of current is much smaller. For example, in a computer's microprocessor chip, a few nanoamperes will suffice for the execution of complicated, rapid electronic processes.

Resistance

We define electrical *resistance* as the opposition that a component, device, or circuit offers to the flow of electric current. The standard unit of resistance is the *ohm*, symbolized Ω in some texts (although we'll write it out in full here as "ohm" or "ohms"). Other common resistance units include the *kilohm* (abbreviated kΩ or k), where

$$1 \text{ k} = 1000 \text{ ohms}$$

and the *megohm* (abbreviated MΩ or M), where

$$1 \text{ M} = 1000 \text{ k}$$
$$= 10^6 \text{ ohms}$$

In the "real world," we'll never encounter any physical object that completely lacks resistance. We call materials with low resistance *conductors*. Elemental silver is an excellent conductor; so are copper and aluminum. In some electronic applications, engineers select circuit materials on the basis of their resistance, with the notion "the larger the better." Materials with exceptionally high resistance constitute *insulators* or *dielectrics*. Glass, plastic, and dry air constitute examples of excellent insulators.

TIP *In a practical electronic circuit, the resistance of a particular component might vary depending on the conditions to which we subject it. A transistor, for example, might have high resistance some of the time, and low resistance at other times. The high/low resistance fluctuation in a transistor can take place thousands, millions, or billions of times each second.*

Still Struggling

The expressions nano- (n), micro- (μ), milli- (m), kilo- (k), and mega- (M) constitute *prefix multipliers* used by scientists to denote large and small quantities. Table 1-1 shows a complete list of prefix multipliers. We base this system on orders of magnitude, or powers of 10. Some computer scientists use another prefix-multiplier system based on powers of 2. They're a little different from the ones shown here..

Electromotive Force

The standard unit of *electromotive force* (EMF) is the *volt*. Engineers abbreviate the word "volt" or "volts" as V. An EMF of 1 volt (1 V) across a component with a resistance of 1 ohm drives a current of 1A through that component. If the term "electromotive force" seems esoteric, think of it as "electrical pressure." Most people call it *voltage*.

Common voltage units on the small scale, besides the volt itself, include the *microvolt* (μV), where

$$1 \ \mu V = 0.000001 \ V$$
$$= 10^{-6} \ V$$

and the *millivolt* (mV), where

$$1 \ mV = 0.001 \ V$$

On the larger scale, we'll encounter the *kilovolt* (abbreviated kV), where

$$1 \ kV = 1000 \ V$$

and the *megavolt* (abbreviated MV), where

$$1 \ MV = 1000 \ kV$$
$$= 10^6 \ V$$

We'll often encounter situations where a significant EMF exists without any flow of current. Lay people call this condition *static electricity*. In this context, "static" means "stationary." Static electricity builds up to fantastic voltages in thundershowers immediately before a *lightning stroke* occurs. Static electricity

Prefix	Symbol	Multiplier	Exponent
TABLE 1-1 Prefix multipliers and their abbreviations.			
yocto–	y	× 0.000000000000000000000001	(10^{-24})
zepto–	z	× 0.000000000000000000001	(10^{-21})
atto–	a	× 0.000000000000000001	(10^{-18})
femto–	f	× 0.000000000000001	(10^{-15})
pico–	p	× 0.000000000001	(10^{-12})
nano–	n	× 0.000000001	(10^{-9})
micro–	m	× 0.000001	(10^{-6})
milli–	m	× 0.001	
centi–	c	× 0.01	
deci–	d	× 0.1	
(none)	--	× 1	
deka–	da or D	× 10	
hecto–	h	× 100	
kilo–	k	× 1000	
mega–	M	× 1,000,000	(10^{6})
giga–	G	× 1,000,000,000	(10^{9})
tera–	T	× 1,000,000,000,000	(10^{12})
peta–	P	× 1,000,000,000,000,000	(10^{15})
exa–	E	× 1,000,000,000,000,000,000	(10^{18})
zetta–	Z	× 1,000,000,000,000,000,000,000	(10^{21})
yotta–	Y	× 1,000,000,000,000,000,000,000,000	(10^{24})

also exists across the terminals of a battery when we don't connect it to anything.

Charge carriers can move only if we provide a conductive path for them to follow. The mere fact that we observe a large EMF between two points does not guarantee that we will observe a large current—or any current.

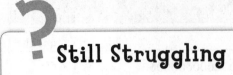

Still Struggling

Have you heard the statement "It's the current, not the voltage, that kills"? Technically that's true, but it doesn't tell the whole story. The current gives you the shock, but it couldn't exist without a source of EMF to make it flow. If a voltage source can move many coulombs of charge carriers in a short time, even a moderate voltage, such as 25 V or 50 V, can drive a deadly current through your body.

Pure versus Variable DC

Common dry cells, "lantern" batteries, automotive batteries, photovoltaic (PV) cells, and computer power packs produce constant DC voltage known as *pure DC*. We can portray a pure DC voltage as a straight, horizontal line on a graph of voltage versus time as shown in Fig. 1-1A on page 4. In some instances a DC voltage pulsates or oscillates rapidly with time; we see an example in Fig. 1-1B. In this situation, we have *pulsating DC*, or DC with an *AC component* superimposed. Once in awhile, we'll encounter a DC voltage that varies in an irregular way as shown in Fig. 1-1C. If we ever observe a *polarity shift* (as we do in the case of Fig. 1-1D), however, we can't call the phenomenon DC.

TIP   *This book contains diagrams of simple electrical and electronic circuits. When we want to show individual components (such as resistors, batteries, capacitors, coils, and transistors) in a* circuit diagram *or* schematic diagram, *we must use special symbols to represent them. If you haven't worked with schematic diagrams before, they might confuse you. In order to get familiar with the symbols, spend some time studying the Appendix at the end of this book.*

Ohm's Law

We can simplify any DC circuit to one source of voltage, a specific current drawn from that source, and a single overall resistance, symbolized as follows:

- To represent the voltage, we write an italic uppercase letter E
- To represent the current, we write an italic talic uppercase letter I
- To represent the resistance, we write an uppercase italic R

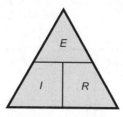

FIGURE 1-2 • The "*EIR* triangle" can serve as a memory aid.

These quantities relate in an orderly fashion called *Ohm's Law*, named after the German physicist *Georg Simon Ohm* (1789–1854). Three formulas denote Ohm's Law:

$$E = IR$$

$$I = E/R$$

$$R = E/I$$

Figure 1-2 shows E, I, and R in a triangle, going in alphabetical order from top to bottom and left to right. If you want to find the formula for any quantity in terms of the other two, cover up the variable for the parameter that you want to find. Then look at the way the other two appear together.

TIP *If the initial quantities are given in units other than volts, azmperes, and ohms, you must convert the voltage, current, and resistance to standard units and then do the arithmetic. After that, you can convert the answer into whatever unit you like.*

Current, Voltage, and Resistance Calculations

Figure 1-3 shows a variable-voltage source or *DC generator*, a *voltmeter* called V for measuring EMF, an *ammeter* called A for measuring current, and a variable resistance R provided by an adjustable resistor called a *potentiometer*. The following three problems (and solutions) demonstrate how we can use Ohm's Law to calculate unknown current I, voltage E, and resistance R in a DC circuit wired as shown in Fig. 1-3.

PROBLEM **1-1**

Suppose that the DC generator produces 10 V, and you set the potentiometer to a resistance of 10 ohms. What's the current?

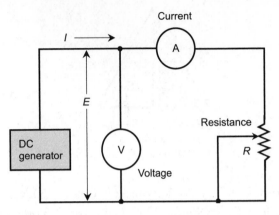

FIGURE 1-3 · Circuit for demonstrating Ohm's Law.

☑ SOLUTION

Use the Ohm's Law formula for obtaining current in terms of voltage and resistance. Because you know the voltage in volts and the resistance in ohms, you can directly plug in the values $E = 10$ and $R = 10$ to get

$$I = E/R$$
$$= 10/10$$
$$= 1.0 \text{ A}$$

PROBLEM **1-2**

Suppose that you set the potentiometer in the circuit of Fig. 1-3 to 100 ohms and measure the current as 10 mA. How much voltage appears across the potentiometer?

☑ SOLUTION

Use the formula for voltage in terms of current and resistance. First, convert the current to amperes: 10 mA = 0.010 A. You already know the resistance in ohms, so you can multiply the current by the resistance to obtain

$$E = IR$$
$$= 0.010 \times 100$$
$$= 1.0 \text{ V}$$

PROBLEM 1-3

If the voltmeter in the circuit of Fig. 1-3 reads 24 V and the ammeter shows 3.0 A, what's the resistance of the potentiometer?

SOLUTION

Use the formula for resistance in terms of voltage and current. You know the voltage in volts and the current in amperes, so you don't have to do any unit conversions. You can plug in E = 24 and I = 3.0 directly, obtaining

$$R = E/I$$
$$= 24/3.0$$
$$= 8.0 \text{ ohms}$$

Power Calculations

If we know at least two of the parameters E, I, and R in a DC circuit, such as the one shown in Fig. 1-3, we can calculate the electrical *power* dissipated in the resistance. The standard unit of power is the *watt*, symbolized by the uppercase nonitalic letter W. When we want to express power as a variable in an equation, we use the uppercase italic letter P. If we know the voltage and the current, then

$$P = EI$$

If we know the current and the resistance, then

$$P = I^2R$$

If we know the voltage and the resistance, then

$$P = E^2/R$$

The foregoing three formulas hold true for units of power in watts, voltage in volts, current in amperes, and resistance in ohms.

TIP   *If you use nonstandard units for any of the input parameters (such as voltage in millivolts and resistance in kilohms, or current in microamperes and voltage in kilovolts), then you must convert the inputs to standard units (volts, amperes, and ohms) before you perform any calculations.*

PROBLEM 1-4

Suppose that the voltmeter in Fig. 1-3 reads 12 V and the ammeter shows 50 mA. Calculate the power dissipated in the potentiometer.

SOLUTION

First, convert the current to amperes, getting $I = 0.050$ A. You know the voltage directly in volts as $E = 12$ V. Therefore

$$P = EI$$
$$= 12 \times 0.050$$
$$= 0.60 \text{ W}$$

PROBLEM 1-5

If the resistance in the circuit of Fig. 1-3 equals 999 ohms and the voltage source delivers 3.0 V, how much power does the potentiometer dissipate?

SOLUTION

You know the resistance directly in ohms and the voltage directly in volts, so you need not do any unit conversions. You can calculate the power simply as

$$P = E^2/R$$
$$= 3.0^2/999$$
$$= 9.0/999$$
$$= 0.0090 \text{ W}$$

You can express this result as 9.0 milliwatts (mW), remembering that 1 mW equals precisely 0.001 W.

PROBLEM 1-6

If the resistance in the circuit of Fig. 1-3 equals 0.47 ohms and the current equals 680 mA, find the power dissipated in the potentiometer.

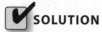**SOLUTION**

First, convert the current to amperes, getting $I = 0.680$ A. You already know resistance in ohms, so you don't have to convert it. You should get

$$P = I^2 R$$
$$= 0.680^2 \times 0.47$$
$$= 0.22 \text{ W}$$

Resistive Networks

Engineers often use combinations of two or more resistors to regulate the current and/or control the voltage in DC circuits. When we interconnect two or more resistors to get special properties, we have a *resistive network*.

Resistances in Series

When we place two or more resistors in *series* (end-to-end like the links in a chain), their values add up arithmetically to yield the total resistance or *net resistance*, as long as we express all of the resistances in the same units. If we express some of the resistances in kilohms and others in megohms, we must convert all the resistances to the same units (ohms, kilohms, or megohms) before adding them up.

Resistances in Parallel

When we connect resistances in *parallel* (across each other like the rungs in a ladder), they behave differently than they do in a series combination. In general, if we have a resistor of a certain value and we place other resistors in parallel with it, the net resistance decreases. The net resistance always has a value smaller than the lowest individual resistance.

Let's imagine resistors as conductive components, or *conductances*. Every component that *impedes* the flow of current also *allows* current to flow; the two parameters occur in inverse proportion. As the resistance goes up, the conductance goes down and vice-versa. The symbol for conductance is an italic uppercase letter G. The standard unit of conductance is the *siemens*, symbolized as a nonitalic uppercase letter S. The conductance G of a component, in siemens (not siemenses!—the plural of this term is the same as the singular), relates to the resistance R, in ohms, according to the formulas

$$G = 1/R$$

and

$$R = 1/G$$

If we connect two or more resistors in parallel, then their *conductance values* add up directly. If we change all the ohmic values to siemens, we can add the numbers and then convert the result back to ohms. The following steps describe how we can figure out the net resistance of a set of three or more resistances in parallel:

- Convert all resistance values to ohms
- Find the reciprocal of each resistance to get the respective conductances in siemens
- Add up the individual conductances to obtain the net conductance
- Take the reciprocal of the net conductance in siemens to derive the net resistance in ohms

TIP   *When you find several resistors connected in parallel and they all have the same resistance, then the net resistance of the combination equals the resistance of any one component divided by the number of components.*

PROBLEM 1-7

Imagine that you connect three resistors of the following values in series with each other: $R_1 = 112$ ohms, $R_2 = 470$ ohms, and $R_3 = 680$ ohms. What's the net series resistance, R, of this combination in kilohms? Round the answer off to the nearest hundredth of a kilohm.

SOLUTION

You can calculate the net resistance, as determined across the whole series combination, by adding the values to obtain

$$R = R_1 + R_2 + R_3$$
$$= 112 + 470 + 680$$
$$= 1262 \text{ ohms}$$

To convert this resistance value to kilohms, you can divide by 1000 to get 1.262 k, which rounds off to 1.26 k.

PROBLEM 1-8

Consider five resistors in parallel. Call them R_1 through R_5, and call the total resistance R as shown in Fig. 1-4. Let R_1 = 100 ohms, R_2 = 200 ohms, R_3 = 300 ohms, R_4 = 400 ohms, and R_5 = 500 ohms. What's the net parallel resistance, R? Round off the answer to the nearest tenth of an ohm.

SOLUTION

To begin, convert the resistances to conductances by taking their reciprocals, as follows:

$$G_1 = 1/100$$
$$= 0.01 \text{ S}$$

$$G_2 = 1/200$$
$$= 0.005 \text{ S}$$

$$G_3 = 1/300$$
$$= 0.00333 \text{ S}$$

$$G_4 = 1/400$$
$$= 0.0025 \text{ S}$$

$$G_5 = 1/500$$
$$= 0.002 \text{ S}$$

When you add all of these conductance values up, you get

$$G = 0.01 + 0.005 + 0.00333 + 0.0025 + 0.002$$
$$= 0.02283 \text{ S}$$

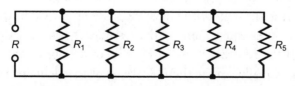

FIGURE 1-4 • Five resistances in parallel, R_1 through R_5, connect like "rungs on a ladder" and yield a total resistance R.

The total resistance, rounded off to the nearest tenth of an ohm, is therefore

$$R = 1/G$$
$$= 1/0.02283$$
$$= 43.8 \text{ ohms}$$

Still Struggling

You can take advantage of two formulas to directly calculate the net resistance of a parallel combination. You don't have to memorize these formulas (but you'll suffer no harm if you do). For two resistances R_1 and R_2 in parallel, you can find the net resistance R as

$$R = R_1 R_2 / (R_1 + R_2)$$

For *n* resistances $R_1, R_2, R_3, ...,$ and R_n in parallel where *n* can represent any positive whole number, calculate

$$R = 1/(1/R_1 + 1/R_2 + 1/R_3 + ... + 1/R_n)$$

Remember to convert all resistances to ohms before you do any arithmetic.

Division of Power

When we connect combinations of resistors to a single source of voltage, the resistors each draw some current from that source. We can determine the total current by calculating the total resistance of the combination, considering the whole network as a single resistor, and then applying Ohm's Law to get the current in terms of the source voltage and the net resistance.

If the resistors in the network all have the same ohmic value, the power from the source distributes equally among the resistances, whether they're in series or in parallel. If the resistors don't all have identical ohmic values, the power won't divide up equally among them. If we want to calculate the power dissipated by any particular resistor in a case of that sort, we must determine either the current through the individual resistor or the voltage across it, and then use the appropriate DC power formula.

Resistances in Series-Parallel

We can connect a group of resistors, all having identical ohmic values, together in parallel sets of series components, or in series sets of parallel components. We call such circuits *series-parallel resistive networks*. When we combine numerous identical resistors together this way, we can greatly increase the total power-handling capacity over that of a single resistor.

The total resistance of a series-parallel resistor network equals the value of any individual resistor if it meets two criteria:

- All of the resistors have the same ohmic value
- The resistors form an *n-by-n matrix*

An *n*-by-*n* matrix is a symmetrical network of components such that, if we let *n* equal some whole number, the circuit has *n* parallel-connected sets of *n* series-connected components (Fig. 1-5A) or else *n* series-connected sets of *n* parallel-connected components (Fig. 1-5B). These two arrangements, although geometrically different, yield identical results in practice. In both of the situations shown here, *n* = 3 so we have a 3-by-3 matrix.

Engineers and technicians sometimes build *n-by-n* matrices to obtain resistive components with large power-handling capacity. In such a circuit, all the resistors must have the same power-dissipation rating. In that case, the combination of *n* by *n* resistors will have n^2 times the power-dissipation capability of

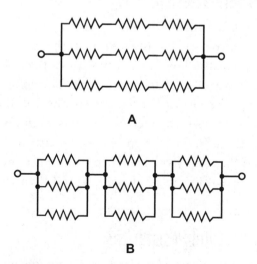

A

B

FIGURE 1-5 · Series-parallel combinations. At A, sets of series resistors connected in parallel. At B, sets of parallel resistors connected in series.

a single resistor. For example, a 3-by-3 series-parallel matrix of 2-W resistors can handle $3^2 \times 2 = 18$ W, and a 10-by-10 array of 1-W resistors can dissipate $10^2 \times 1 = 100$ W.

TIP *Nonsymmetrical series-parallel networks assembled from identical resistors can increase the power-handling capability over that of a single resistor. But in situations of this sort, for example a 5-by-7 matrix or an 8-by-3 matrix, the total resistance differs from the value of any individual resistor. Nevertheless, the over-all power-handling capacity multiplies by the total number of resistors whether the network is symmetrical or not, provided that all the resistors have the same resistance and the same power-handling capacity.*

Currents through Series Resistances

In a series DC circuit, the current at any particular point equals the current at any other point. This rule holds true no matter what the components actually are, and regardless of whether or not they all have the same resistance. If the various components have different resistances, then some of them consume more power than others. In case one of the components shorts out, the current through the whole chain increases because the overall resistance of the chain goes down. If a component opens, the current drops to zero at every point because charge carriers can't flow.

Voltages across Series Resistances

In a series DC circuit, the voltage divides up among the components. The sum total of the voltages across each resistance equals the supply voltage. This rule always holds true, no matter how large or how small the resistances might get, and whether or not they all have the same value. The voltage across any individual resistor equals the product of the current through that resistor and its ohmic value, as long as we remember to use volts, ohms, and amperes when making our calculations. To find the current in the circuit, we must know the total resistance and the supply voltage. Then we can use the formula $I = E/R$. First we find the current in the whole circuit, and then we find the voltage across the resistor of interest.

Voltages across Parallel Resistances

In a parallel DC circuit, the voltage across each component equals the supply or battery voltage. The current drawn by any particular component depends on its individual resistance. In this sense, the components in a parallel-wired circuit work independently, as opposed to the series-wired circuit in which they inter-act. If we remove any particular branch from a parallel circuit (simply pull it out, leaving the ladder with a "missing rung"), the conditions in the other branches remain the same as they were before we made the change. If we add a new branch, assuming the power supply can deliver enough current, condi-tions don't change in any of the original branches.

TIP   *When a DC voltage exists across a particular component or set of compo-nents, or between two specific points in a circuit, engineers sometimes refer to that voltage as a* **potential difference.**

Currents through Parallel Resistances

Figure 1-6 shows a circuit containing several resistors connected in parallel across a DC battery. Let R represent the net resistance of the entire parallel combina-tion. Let E represent the battery voltage. Suppose that we measure the current in some branch n, containing the resistance we call R_n, using ammeter A as shown. Let I_n represent this current. Ohm's Law tells us that

$$I_n = E/R_n$$

Now imagine that we measure the individual currents through all of the other branches, one at a time. Finally, we add up all of the currents in the individual

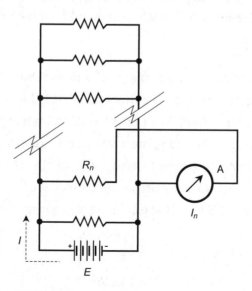

FIGURE 1-6 · Analysis of current in a parallel DC circuit.

branches. Invariably, we'll find that the sum of the currents through all the resistances equals the total current, I, drawn from the battery.

PROBLEM 1-9

Consider a specific case of the scenario shown in Fig. 1-6. Imagine that you have 10 resistors connected in parallel. One of these resistors, R_n, has a value of 100 ohms. Suppose that the current through R_n equals 0.173 A. What's the voltage, E, of the battery?

SOLUTION

You should begin by finding the voltage across R_n. In a parallel DC circuit, the entire supply or battery voltage appears across each individual resistor. You therefore have a simple case of Ohm's Law, where

$$E = IR$$
$$= 0.173 \times 100$$
$$= 17.3\,\text{V}$$

PROBLEM 1-10

Suppose that in the situation described in Problem 1-9, all 10 of the resistors have values of 100 ohms. If the battery supplies 17.3V, how much current does the entire parallel combination of resistors draw? How much power does the entire combination demand from the battery?

SOLUTION

The total current equals the sum of the currents through each of the resistors. We already know that this current equals 0.173 A in the case of R_n. Because all the resistors have the same value, the current through each resistor must equal 0.173 A. Therefore, the total current equals $0.173 \times 10 = 1.73$ A. We can determine the power P demanded from the battery using the formula for power in terms of voltage and current. We know that the battery voltage E equals 17.3 V and the total current I equals 1.73 A, so

$$P = EI$$
$$= 17.3 \times 1.73$$
$$= 29.9\,\text{W}$$

Power Distribution in Series Circuits

When we want to find the power dissipated by a particular resistor R_n in a circuit containing n resistors in series, we can first determine the current, I, that the circuit carries. Then we can easily calculate the power P_n, based on the formula

$$P_n = I^2 R_n$$

The total power, or *wattage*, dissipated in a series circuit equals the sum of the wattages dissipated in each resistor.

Power Distribution in Parallel Circuits

When we connect two or more resistors of various ohmic values in parallel, the current differs in each component, but we can find the power P_n, dissipated by a particular resistor R_n, using the formula

$$P_n = E^2 / R_n$$

where E represents the voltage of the battery or power supply. In a parallel circuit, the voltage across any particular resistor equals the voltage across any other resistor, and the total dissipated wattage equals the sum of the wattages dissipated in the individual resistors.

Kirchhoff's Laws

In all DC circuits, the currents and voltages behave according to two physical principles known as *Kirchhoff's laws*. Let's name them according to their descriptions of current and voltage, as follows.

- According to *Kirchhoff's Law for Current*, the current entering any point in a multiple-branch DC circuit equals the current emerging from that point, no matter how many branches lead into or out of the point (Fig. 1-7A).
- According to *Kirchhoff's Law for Voltage*, the sum of all the voltages starting at a fixed point in a DC circuit, proceeding all the way around and returning to that point from the opposite direction, and taking polarity into account, equals zero (Fig. 1-7B).

TIP   *The current law demonstrates the fact that in a DC circuit, current can't come from nowhere or disappear into nowhere; we always observe "conservation of current." The voltage rule holds true because in a DC circuit, a potential difference can't from nowhere or disappear into nowhere; we always observe "conservation of voltage."*

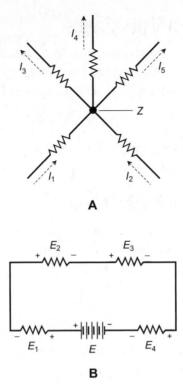

A

B

FIGURE 1-7 • At A, Kirchhoff's Law for Current. The current entering point Z equals the current leaving point Z. In this case, $I_1 + I_2 = I_3 + I_4 + I_5$. At B, Kirchhoff's Law for Voltage. In this case, $E + E_1 + E_2 + E_3 + E_4 = 0$, taking polarity into account.

Voltage Dividers

When we connect resistors in series and then apply a potential difference across the whole combination, we obtain various voltages across the individual components. By carefully choosing ohmic values for the resistors, we can "tailor" those voltages to meet certain needs. A circuit of series-connected resistances, deliberately set up to produce specific DC voltages at various points, is called a *voltage divider*.

In the construction of a voltage divider, we should make the resistances as small as possible without draining excessive current from the power supply.

This practice ensures that the voltages across the individual resistors won't fluctuate too much under varying external circuit conditions. Any system that is connected to a voltage divider and relies on that divider for its operation affects the voltages across the divider resistors. As we make the divider resistances smaller, the extent of this undesirable voltage fluctuation decreases. Ideally, the total series resistance of the divider should constitute only a small fraction of the *load resistance* (the resistance of the external system operating from the divider).

Figure 1-8 illustrates the principle of voltage division. To find the total resistance R, we add up all the individual resistances to obtain

$$R = R_1 + R_2 + R_3 + \ldots + R_n$$

because the resistors all appear in series. The battery voltage equals E, so the current drawn from the battery is

$$I = E/R$$

At the points $P_1, P_2, P_3, \ldots, P_n$, the voltages, with respect to the negative terminal of the battery, equal $E_1, E_2, E_3, \ldots, E_n$. The highest voltage, E_n, equals the supply voltage, E. Each and every one of the other voltages is less than E; if we measure them, we'll see that

$$E_1 < E_2 < E_3 < \ldots < E_n$$

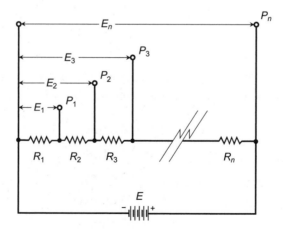

FIGURE 1-8 · General arrangement for a voltage-divider circuit.

The voltages at the various points increase according to the sum total of the resistances up to each point, in proportion to the total resistance, multiplied by the supply voltage. We have the voltage sequence

$$E_1 = E \, (R_1/R)$$
$$E_2 = E \, (R_1 + R_2)/R$$
$$E_3 = E \, (R_1 + R_2 + R_3)/R$$
$$\downarrow$$
$$E_n = E \, (R_1 + R_2 + R_3 + \ldots + R_n)/R$$
$$= E \, (R/R)$$
$$= E$$

We can also calculate these voltages if we multiply the net circuit current I by the incremental resistances one by one, obtaining the sequence

$$E_1 = IR_1$$
$$E_2 = I \, (R_1 + R_2)$$
$$E_3 = I \, (R_1 + R_2 + R_3)$$
$$\downarrow$$
$$E_n = I \, (R_1 + R_2 + R_3 + \ldots + R_n)$$
$$= IR$$
$$= E$$

PROBLEM 1-11

Consider a voltage divider constructed according to the circuit design shown in Fig. 1-8, in which we have five resistors with the following values:

$$R_1 = 300 \text{ ohms}$$
$$R_2 = 250 \text{ ohms}$$
$$R_3 = 200 \text{ ohms}$$
$$R_4 = 150 \text{ ohms}$$
$$R_5 = 100 \text{ ohms}$$

If the battery voltage E equals 100 V, what are the divider voltages E_1 through E_5 at the points P_1 through P_5?

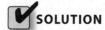SOLUTION

We can calculate the total resistance R of the series combination as

$$R = R_1 + R_2 + R_3 + R_4 + R_5$$
$$= 300 + 250 + 200 + 150 + 100$$
$$= 1000 \text{ ohms}$$

Therefore, the current I through the circuit is

$$I = E/R$$
$$= 100/1000$$
$$= 0.100 \text{ A}$$
$$= 100 \text{ mA}$$

This current flows through each of the resistors. We can determine the voltages E_1 through E_5 if we multiply I by the incremental resistances at the points P_1 through P_5 (the total resistances between those points and the negative battery terminal) to get the following values:

$$E_1 = 0.100 \times 300$$
$$= 30.0 \text{ V}$$

$$E_2 = 0.100 \times (300 + 250)$$
$$= 0.100 \times 550$$
$$= 55.0 \text{ V}$$

$$E_3 = 0.100 \times (300 + 250 + 200)$$
$$= 0.100 \times 750$$
$$= 75.0 \text{ V}$$

$$E_4 = 0.100 \times (300 + 250 + 200 + 150)$$
$$= 0.100 \times 900$$
$$= 90.0 \text{ V}$$

$$E_5 = 0.100 \times (300 + 250 + 200 + 150 + 100)$$
$$= 0.100 \times 1000$$
$$= 100 \text{ V}$$

PROBLEM 1-12

How much power, in watts, does each resistor dissipate in the situation of Prob. 1-11 and its solution?

✔ SOLUTION

Let's call these power levels P_1 through P_5. (In this context, P refers to the word "power," not a "point" as in Fig. 1-8!) We observe the same current I through each of the resistors, and we know that $I = 0.100$ A. The square of the current, I^2, equals 0.0100. Now we can calculate the power values one at a time, as follows:

$$P_1 = I^2 R_1$$
$$= 0.0100 \times 300$$
$$= 3.00 \text{ W}$$

$$P_2 = I^2 R_2$$
$$= 0.0100 \times 250$$
$$= 2.50 \text{ W}$$

$$P_3 = I^2 R_3$$
$$= 0.0100 \times 200$$
$$= 2.00 \text{ W}$$

$$P_4 = I^2 R_4$$
$$= 0.0100 \times 150$$
$$= 1.50 \text{ W}$$

$$P_5 = I^2 R_5$$
$$= 0.0100 \times 100$$
$$= 1.00 \text{ W}$$

DC Magnetism

Whenever the atoms in a sample of matter align themselves to some extent, a *magnetic field* exists. A magnetic field can also result from the motion of electric charge carriers such as electrons through any material substance, or even through *free space* (a vacuum).

Lines of Flux

We can imagine a magnetic field as *flux lines* or *lines of flux*, where each individual "line" represents a certain quantity of magnetism. We can express the intensity of a magnetic field, also known as the *flux density*, in terms of the number of flux lines passing through a surface or flat region having a specific area, such as a square centimeter (cm^2) or a square meter (m^2). In calculations and equations, engineers symbolize flux density by writing an uppercase italic letter *B*. The standard unit of magnetic flux density is the *tesla*, abbreviated as a nonitalic uppercase letter T.

Scientists consider magnetic fields to originate at *north poles* and terminate at *south poles*. In the vicinity of a magnetic object, all the flux lines emerge from the north pole and converge toward the south pole. Every line of flux connects the two poles. Around a bar or rod-shaped magnet, the greatest flux density exists near the poles where the lines diverge or converge. Around a current-carrying wire, the greatest flux density exists close to the wire. We can always follow a flux line around a *closed curve*, starting at some point, moving away from it in one direction, and finally arriving back again from the opposite direction.

TIP   *A flux line isn't a material object, but a geometric abstraction. In addition, the term "line" can easily mislead us because, in the overall sense, flux lines usually manifest themselves as curves, not straight paths.*

Magnetomotive Force

When we work with magnetic fields produced by electric currents, we can quantify a phenomenon called *magnetomotive force* with a unit called the *ampere-turn* (At). This unit describes itself well. Magnetomotive force in ampere-turns equals the number of amperes flowing in a loop or coil of wire, times the number of turns that the loop or coil contains.

If we bend a length of wire into a loop and drive 1 A of current through it, we get 1 At of magnetomotive force in the vicinity of the loop. If we wind the same length of wire (or any other length) into a 50-turn coil and keep driving 1 A of current through it, the resulting magnetomotive force goes up to 50 At. If we then reduce the current in the 50-turn loop to 1/50 A or 20 mA, the magnetomotive force goes back down to 1 At.

Sometimes, engineers employ a unit called the *gilbert* to express magnetomotive force. One gilbert (1 Gb) equals exactly $5/(2\pi)$, or approximately 0.7958 At. This relation yields conversion algorithms as follows:

- If we want to determine the number of ampere-turns when we know the number of gilberts, we should multiply by 5/(2π) or approximately 0.7958.
- If we want to determine the number of gilberts when we know the number of ampere-turns, we should multiply by 2π/5 or approximately 1.257.

Flux Density versus Current

In a straight wire carrying a steady, direct electric current and surrounded by free space or air, we observe the greatest flux density near the wire, and diminishing flux density as we get farther away from the wire. We can use a simple formula to express magnetic flux density as a function of the current in a straight wire and the distance from the wire. Like many physics and engineering formulas, it works perfectly under ideal circumstances, but discrepancies often occur in the "real world."

Imagine an infinitely thin, absolutely straight, infinitely long length of wire (the ideal case). Suppose that the wire carries a current of I amperes. Let's represent the flux density (in teslas) as B. Now let's consider a point P that lies at a distance r (in meters) from the wire. We can find the flux density at the point P using the formula

$$B = 2 \times 10^{-7} \, I/r$$

We can consider the value of the constant, 2×10^{-7}, mathematically exact to any desired number of significant figures.

TIP   *Of course, we'll never encounter a wire with zero thickness or infinite length in the "real world." But as long as the wire thickness constitutes a small fraction of r, and as long as the wire lies along a reasonably straight line near point P, the foregoing formula works okay in most practical scenarios.*

Ampere's Law

At any fixed point near an electrical conductor carrying DC, the magnetic-field intensity varies in direct proportion to the current in the conductor. For a straight wire, the magnetic flux lines take the form of concentric circles centered on the wire. Figure 1-9 shows a cross-sectional view of this situation. Imagine the wire running perpendicular to the page, so it appears as a point in this diagram.

In a DC circuit, physicists define *theoretical current*, also called *conventional current*, as flowing from the positive electric pole to the negative electric pole—even though electrons move from negative to positive. With this convention in

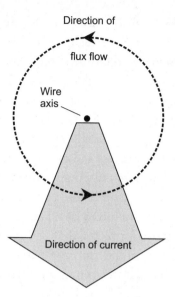

FIGURE 1-9 • Ampere's Law for a current-carrying conductor.

mind, imagine that the theoretical current in the situation of Fig. 1-9 flows along the wire, straight out of the page toward you. According to *Ampere's Law*, the magnetic flux circulates counterclockwise as you see it, going around circles centered on the wire axis.

Still Struggling

Ampere's Law is sometimes called the *right-hand rule.* If you hold your right hand with the thumb pointing out straight and the fingers curled, and then point your thumb in the direction of the positive-to-negative theoretical current in a straight wire, your fingers curl in the direction of the magnetic flux flow. Similarly, if you orient your right hand so your fingers curl in the direction of the magnetic flux circulation, your thumb points in the direction of the theoretical current flow.

Permeability

Permeability quantifies the way in which a substance affects magnetic flux density. Engineers use the term *ferromagnetic* to describe any material that concentrates magnetic lines of flux compared with their concentration in free space. We can "magnetize" samples of such substances. Iron provides the most well-known example. As the permeability of a sample increases, so does the extent to which it concentrates magnetic lines of flux. A substance that dilates, or dilutes, magnetic flux is called *diamagnetic*. Wax, dry wood, bismuth, and silver exhibit this property. The dilution of magnetic flux in a diamagnetic medium never occurs to a great extent, but ferromagnetic materials can concentrate flux lines tremendously. Physicists and engineers express and measure magnetic permeability on a scale relative to free space, which has a permeability of exactly 1 by definition.

Suppose that you place a rod-shaped iron core inside a long, cylindrical coil (called a *solenoidal coil*) with the intent of making an *electromagnet*. The flux density in the core increases by a factor that can range from about 60 up to 8000 (depending on the purity of the iron) compared with the flux density that would exist inside the same coil if the iron weren't there. Therefore, the permeability of iron can range from roughly 60 to 8000. If you use a specially formulated high-permeability metal alloy as the core material in an electromagnet, you can increase the flux density inside the core by as much as 1,000,000 times. Table 1-2 lists approximate permeability factors for a few common materials.

TABLE 1-2 Magnetic permeability factors for some common materials. All figures are approximate, with the exception of the value for free space (a vacuum).

Substance	Permeability
Air, dry	Slightly more than 1
Aluminum	Slightly more than 1
Bismuth	Slightly less than 1
Cobalt	60 to 70
Ferrite	100 to 3000
Free space (vacuum)	1 (exactly)
Iron	60 to 8000
Nickel	50 to 60
Permalloy	3000 to 30,000
Silver	Slightly less than 1
Steel	300 to 600
Specialized alloys	100,000 to 1,000,000
Wax	Slightly less than 1
Wood, dry	Slightly less than 1

Diamagnetic substances can keep magnetic objects physically separated while minimizing the interaction between them. Nonferromagnetic metals, such as copper and aluminum, conduct electric current well, but magnetic flux poorly. Sheets of such metals can provide *electrostatic shielding*, a means of allowing magnetic fields to pass through while blocking electric fields. A sheet or screen made of copper or aluminum, connected to a good electrical ground, effectively "shorts out" *electric flux*, but magnetic flux can pass through as if the barrier did not exist.

The Relay

A *relay* (Fig. 1-10) employs an electromagnet to allow remote-control switching. A spring (not shown) holds a movable lever, called the *armature*, to one side when no current flows through the electromagnet. Under these conditions, terminal X connects to terminal Y, but not to terminal Z. When we apply a sufficient control voltage to the relay coil, thereby driving current through it, the electromagnet pulls the armature over to the other side, disconnecting terminal X from terminal Y, and connecting X to Z instead.

A *normally closed relay* completes the circuit when no current flows in its electromagnet, and breaks the circuit when current flows. A *normally open relay* works in the opposite fashion, completing the circuit when current flows in the coil, and breaking the circuit when no current flows. Some relays have several sets of contacts, allowing engineers and technicians to switch multiple circuits simultaneously. Some relays can remain in one state (either with current or without) for a long time, while others can change states several times per second.

In practice, a limit exists as to how fast a mechanical relay can switch between states, and the contacts eventually become corroded or wear out. For these

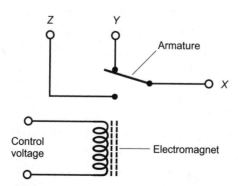

FIGURE 1-10 · Schematic representation of a simple relay.

reasons, relays are rarely used for switching applications these days; semiconductor devices have largely replaced them. However, in some systems, relays still find favor because they can withstand unexpected voltage spikes that semiconductor devices cannot endure. For example, the "surge" caused by a nearby lightning strike might destroy a semiconductor-based switch, while a relay would "shrug off" the event without damage.

Magnetic Tape and Disk

Magnetic tape consists of millions of ferromagnetic particles glued to a flexible, nonmagnetic strip such as plastic. A fluctuating magnetic field, produced by a small electromagnet called the *recording head*, polarizes these particles. As the field changes in strength next to the recording head, the tape passes by at a constant, controlled speed, producing regions in which the particles become polarized in alternating directions. When the tape passes by at the same speed in the playback mode, the magnetic fields around the particles cause a fluctuating field that a tiny coil called the *pickup head* can detect, outputting a variable electric current. The magnetic field produced by the passing tape has the same pattern of variations as the original magnetic field from the recording head.

The principle of the *magnetic disk*, on the micro scale, resembles that of the magnetic tape. The disk stores the information in *binary digital* form; there exist only two different ways that the particles are magnetized. This process results in essentially error-free storage. On a larger scale, the disk works differently than tape because of the difference in geometry. On a tape, the information is physically spread out over a long span, and some *binary digits* (or *bits*) of data lie far away from other bits. But on a disk, no two bits lie farther apart than the diameter of the disk. For this reason, we can "write" and "read" data more quickly to or from a disk than we can do with a tape.

TIP *The data on a magnetic tape or disk can be distorted or erased by external magnetic fields. Extreme heat can also result in data loss and sometimes, even physical damage to the medium. We should never expose magnetic tapes or disks to these adverse conditions.*

PROBLEM 1-13

Suppose that we wind a length of wire exactly 10 times around a plastic hoop with a permeability of 1, comparable to that of free space. Then we connect a battery to the coil so that a current of 0.5000 A flows through it. What's the resulting magnetomotive force in ampere-turns? In gilberts?

SOLUTION

To find the magnetomotive force in ampere-turns, we multiply the current in amperes by the number of coil turns, obtaining $0.5000 \times 10.00 = 5.000$ At. To convert this figure to its equivalent in gilberts, we multiply by 1.257, getting $5.000 \times 1.257 = 6.285$ Gb.

PROBLEM 1-14

What's the flux density B in teslas at a distance of 200 mm from a straight, thin wire carrying 400 mA of DC?

SOLUTION

First, we must convert all quantities to standard units. We therefore have $r = 0.200$ m and $I = 0.400$ A. We can input these values into the formula for flux density to obtain

$$B = 2 \times 10^{-7} I/r$$

$$B = 2.00 \times 10^{-7} \times 0.400/0.200$$

$$= 4.00 \times 10^{-7}\, T$$

QUIZ

Refer to the text in this chapter if necessary. A good score is eight correct. The answers are listed in the back of the book.

1. Figure 1-11 shows three resistors connected to a 12-V battery. Assume that all component values are mathematically exact. How much current flows through the 4.8-ohm resistor?

   A. 0.5 A
   B. 1.0 A
   C. 0.67 A
   D. 1.5 A

2. In the circuit of Fig. 1-11, how much current flows, in total, through the parallel-connected 12-ohm and 18-ohm resistors?

   A. 0.5 A
   B. 1.0 A
   C. 0.67 A
   D. 1.5 A

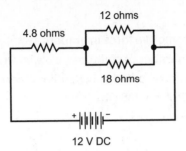

FIGURE 1-11 · Illustration for Quiz
Questions 1 through 7.

3. **In the circuit of Fig. 1-11, how much voltage appears across the 4.8-ohm resistor?**
   A. 2.5 V
   B. 0.4 V
   C. 4.8 V
   D. 4.0 V

4. **In the circuit of Fig. 1-11, how much voltage appears across the 12-ohm resistor?**
   A. 7.2 V
   B. 6.0 V
   C. 4.8 V
   D. 3.6 V

5. **In the circuit of Fig. 1-11, how much current flows through the 18-ohm resistor?**
   A. 1.50 A
   B. 0.72 A
   C. 0.50 A
   D. 0.40 A

6. **In the circuit of Fig. 1-11, how much power does the 4.8-ohm resistor dissipate?**
   A. 0.60 W
   B. 1.2 W
   C. 2.4 W
   D. 4.8 W

7. **In the circuit of Fig. 1-11, how much power does the 18-ohm resistor dissipate?**
   A. 2.88 W
   B. 1.44 W
   C. 0.800 W
   D. 0.720 W

8. **Figure 1-12 shows a wire loop connected to an 18-V battery through a 2.4-ohm resistor. Assume that the component values are exact. How much magneto-motive force does the loop produce?**

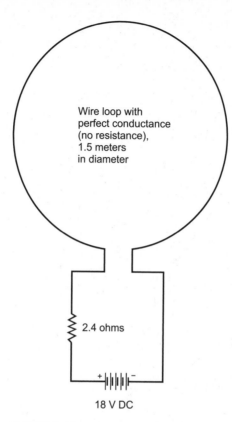

FIGURE 1-12 · Illustration for Quiz Questions 8 through 10.

- A. 0.3333 Gb
- B. 5.967 Gb
- C. 7.500 Gb
- D. 9.428 Gb

9. **If we double the diameter of the loop in the scenario of Fig. 1-12 from 1.5 meters to 3.0 meters, what happens to the magnetomotive force it produces?**
    - A. It decreases to half its former value.
    - B. It increases to twice its former value.
    - C. It increases to four times its former value.
    - D. It stays the same.

10. **If we rearrange the wire in the scenario of Fig. 1-12, winding it into a four-turn coil instead of a single-turn loop without changing the total length of wire, what happens to the magnetomotive force it produces?**
    - A. It decreases to half its former value.
    - B. It increases to twice its former value.
    - C. It increases to four times its former value.
    - D. It stays the same.

chapter 2

Alternating Current

In alternating current (AC), the direction of the current, or the polarity of the voltage, reverses repeatedly. In most AC utility circuits in the United States, 60 complete wave cycles occur every second; the current direction or voltage polarity reverses 120 times per second. In radio transmitters and receivers, an AC wave can reverse polarity thousands or millions of times per second.

CHAPTER OBJECTIVES

In this chapter you will

- Define and quantify AC frequency and period.
- Compare sine, square, sawtooth, complex, and irregular AC waveforms.
- Add DC components to AC waves and analyze the results.
- Learn how to express AC amplitude in root-mean-square (RMS) terms.
- Combine and compare waves that differ in phase.
- Analyze AC wave phase relationships.
- See how utility companies transmit electrical power.

Frequency and Waveform

In a *periodic AC wave*, the current or voltage, as a function of time, cycles back and forth indefinitely. We call the length of time between a specific point in a cycle and the same point in the next cycle the *period* of the wave. Figure 2-1 illustrates two complete cycles of an AC wave. On the vertical (amplitude) axis, points above the time line represent current in one direction (or positive voltage), and points below the time line represent current in the other direction (or negative voltage).

A Wide Range

Theoretically, the period of an AC wave can be anything "greater than zero and less than infinity." In practice, we see AC waves with periods as short as a minuscule fraction of a second, and as long as several years.

If we express the period of an AC wave in seconds (abbreviated s), we denote it by writing an uppercase italicized letter T. We denote the *frequency* of an AC wave, expressed in cycles per second, by writing a lowercase italicized letter f. Mathematically, we can state the relation between period and frequency as

$$f = 1/T$$

or as

$$T = 1/f$$

Engineers use a unit called the *hertz* (abbreviated Hz) to express frequency values. A frequency of 1 Hz represents precisely 1 cycle per second. We can

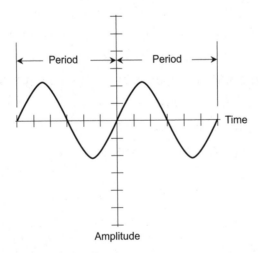

FIGURE 2-1 • A sine wave. The period equals the length of time for one cycle to complete itself.

portray large frequency values in *kilohertz* (kHz), *megahertz* (MHz), *gigahertz* (GHz), or *terahertz* (THz), as follows:

$$1 \text{ kHz} = 1000 \text{ Hz}$$
$$= 10^3 \text{ Hz}$$

$$1 \text{ MHz} = 1000 \text{ kHz}$$
$$= 10^6 \text{ Hz}$$

$$1 \text{ GHz} = 1000 \text{ MHz}$$
$$= 10^9 \text{ Hz}$$

$$1 \text{ THz} = 1000 \text{ GHz}$$
$$= 10^{12} \text{ Hz}$$

In electronics, engineers and technicians work with AC signals in the *audio-frequency* (AF) or *radio-frequency* (RF) ranges. Currents in the AF range have frequencies between 20 Hz and 20 kHz. These figures represent the lowest and highest frequencies, respectively, that a person with "good ears" can hear when we transform AC waves into *sound waves*. Currents in the RF range have frequencies ranging from a few kilohertz up to many gigahertz.

TIP *In the ideal case, an AC wave concentrates all of its energy at a single frequency called the* **fundamental frequency.** *A wave of this type constitutes a* **pure sine wave,** *meaning that if we graph its amplitude as a function of time, we get a rendition of the trigonometric* **sine function.** *In the "real world," an AC wave will likely contain some "impurities" called* **harmonics,** *which exist at whole-number multiples of the fundamental frequency. Some AC signals contain energy at frequencies not clearly related to the fundamental or any of the harmonics. Once in awhile, we'll encounter waves that contain hundreds, thousands, or even infinitely many different* **component frequencies.**

Sine Waves

The classical AC wave has a sine-wave, or *sinusoidal*, shape. Figure 2-1 provides a good example. Any AC wave that concentrates all of its energy at a single frequency has a perfect sine-wave shape. The converse also holds true: A perfect AC sine wave contains only one component frequency—the fundamental.

In practice, a signal can have a waveform that's so close to a sine wave that it looks like the sine function on an *oscilloscope*, when in fact it's far from perfect.

An oscilloscope shows time on the horizontal axis and amplitude (usually voltage) on the vertical axis, so it portrays amplitude as a function of time after the fashion of Fig. 2-1 on page 38. An engineer or technician would say that an oscilloscope has a *time-domain* display.

Utility AC in the United States has an almost perfect sine-wave shape when you look at it on an oscilloscope, but it isn't perfect. The imperfections are responsible for *electromagnetic (EM) noise* that can interfere with radio communications and wireless receivers. If you read my book *Electricity Experiments You Can Do at Home* (McGraw-Hill, 2010), you'll discover how a personal computer can visually display the harmonics and EM noise present on ordinary utility power lines.

Square Waves

On an oscilloscope, a perfect *square wave* looks like a pair of parallel, dashed lines, one having positive polarity and the other having negative polarity as

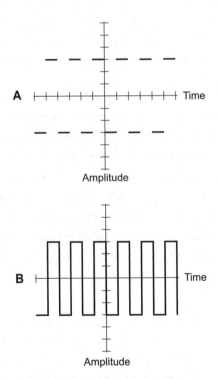

FIGURE 2-2 · At A, a theoretically perfect square wave. At B, the transitions appear as vertical lines for illustrative purposes.

shown in Fig. 2-2A. When we want to draw a graph of a square wave, we can show the instantaneous polarity transitions as vertical lines (Fig. 2-2B).

A square AC wave might have equal negative and positive peaks. In that case, the *absolute amplitude* of the wave remains constant at a certain voltage, current, or power level. Half of the time the amplitude equals a certain positive number of volts or amperes, and the other half of the time the amplitude equals the same number of negative volts or amperes, so we have a *symmetrical square wave*.

In an *asymmetrical square wave*, the positive and negative amplitudes differ. For example, the positive peak voltage might equal +5 V and the negative peak voltage might equal −3 V. It's also possible that the length of time during which the amplitude is positive will differ from the length of time that the amplitude is negative. In a scenario of this sort, we have a so-called *rectangular wave*.

Sawtooth Waves

Besides the sine wave, the square wave, and the rectangular wave, we'll encounter the *sawtooth wave*, which gets its name from its appearance when graphed.

Figure 2-3 illustrates one type of sawtooth wave. The positive-going transition, also called the *attack*, the *leading edge*, or the *rise*, occurs instantly. The negative-going transition, also called the *release*, the *trailing edge*, or the *decay*, occurs gradually. The period equals the time between points at identical positions on successive cycles.

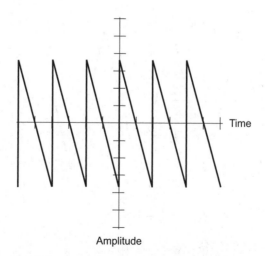

FIGURE 2-3 · A fast-rise, slow-decay sawtooth wave.

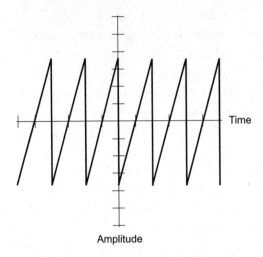

FIGURE 2-4 • A slow-rise, fast-decay sawtooth wave, also called a ramp wave.

Another form of sawtooth wave exhibits a gradual rise and an instantaneous decay. Engineers and technicians call this type of wave, illustrated in Fig. 2-4, a *ramp wave* or simply a *ramp*. Ramp waves produce the beam-scanning currents in old-fashioned cathode-ray-tube (CRT) television sets and oscilloscopes.

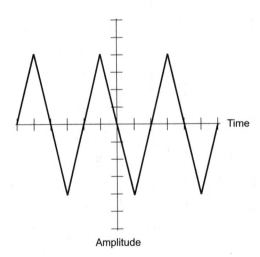

FIGURE 2-5 • A triangular wave.

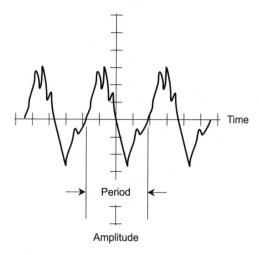

FIGURE 2-6 · A complex waveform.

Sawtooth waves can have rise and decay slopes in an infinite number of different combinations. Figure 2-5 portrays a situation in which the rise and decay have equal, nonzero duration. We call it a *triangular wave*.

Complex Waves

Figure 2-6 shows an example of a *complex wave*, also called an *irregular wave*. We can see that it has a definite period, so it has a specific and measurable frequency. The waveform, although complicated, repeats itself exactly. The period equals the time between points at identical positions on successive repetitions of the waveform.

Frequency-Domain Displays

An AC sine wave, as displayed on a lab instrument called a *spectrum analyzer*, appears as a single vertical line or *pip*, as shown in Fig. 2-7A. The spectrum analyzer shows frequency on the horizontal scale and amplitude on the vertical scale, so it depicts amplitude as a function of frequency. For this reason, engineers call it a *frequency-domain* display.

As we've learned, some AC waves contain energy at harmonics along with energy at the fundamental frequency. If a harmonic wave has a frequency equal to n times the fundamental, then we call that wave the n*th harmonic*.

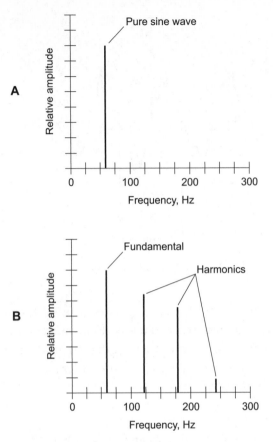

FIGURE 2-7 • At A, a pure 60-Hz sine wave as seen on a spectrum-analyzer display. At B, a 60-Hz wave containing harmonic energy.

In Fig. 2-7B, we see an AC signal's fundamental pip along with several harmonics, as it would look on the display of a spectrum analyzer.

The frequency-domain displays of square waves and sawtooth waves show harmonic components in addition to the energy at the fundamental frequency. The wave shape depends on the amount of energy in the harmonics, and the way in which this energy is distributed among the harmonic frequencies.

Complex waves can produce all sorts of frequency-domain displays. Figure 2-8 shows an example: a spectral display of an amplitude-modulated (AM) voice radio signal. Much of the energy concentrates in the center of the pattern, at the frequency shown by the vertical line. Plenty of energy also exists near, but not exactly at, the center frequency.

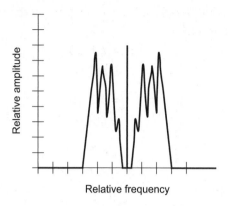

FIGURE 2-8 • An AM radio signal as it might look on a spectrum-analyzer display.

PROBLEM 2-1

One microsecond (symbolized μs) equals 0.000001 of a second (10^{-6} s). What's the frequency, in hertz and in kilohertz, of a wave with a period of 50 μs?

SOLUTION

The frequency f in hertz equals the reciprocal of the period T in seconds. We've been told that T equals 50 μs, which we can also denote as 0.000050 s. Therefore

$$f = 1/0.000050$$

$$= 20,000 \text{ Hz}$$

$$= 20 \text{ kHz}$$

PROBLEM 2-2

What's the frequency of the second harmonic of the wave described in Prob. 2-1? What's the frequency of the third harmonic? What's the frequency of the nth harmonic, where n can represent any positive whole number? Express the answers in kilohertz.

SOLUTION

The fundamental frequency of this wave equals 20 kHz. Let's call the frequency of the second harmonic (in kilohertz) f_2, the frequency of the third harmonic (in kilohertz) f_3, and in general, the frequency of the nth harmonic (in kilohertz) f_n. Now we can portray the second, third, and nth harmonics as follows:

$$f_2 = 2 \times 20$$
$$= 40 \text{ kHz}$$

$$f_3 = 3 \times 20$$
$$= 60 \text{ kHz}$$

$$f_n = n \times 20$$
$$= 20n \text{ kHz}$$

More Definitions

Engineers and scientists break an AC wave cycle down into small parts for analysis and reference. One popular method of describing an AC cycle involves dividing it up into 360 equal *degrees of phase*. We assign the value 0° to the point in the cycle where the amplitude is 0 and positive-going. We give the same point on the next cycle the value 360°. The point halfway through the cycle lies at 180°; a quarter cycle equals 90°. Alternatively, we can divide a wave cycle into 2π (approximately 6.2832) equal parts called *radians of phase*. A radian corresponds to an angle of approximately 57.296°.

Angular Frequency

Sometimes, the frequency of an AC wave is expressed in *degrees per second* rather than in hertz, kilohertz, megahertz, or gigahertz. Because a complete cycle contains 360°, the *angular frequency* of a wave, in degrees per second, equals 360 times the frequency in hertz. We can also express angular frequency in *radians per second* (rad/s). A complete wave cycle contains 2π radians, so the angular frequency of a wave, in radians per second, equals 2π times the frequency in hertz.

Amplitude Expressions

We can express or measure AC wave amplitude in amperes (for current), volts (for voltage), or watts (for power).

The *instantaneous amplitude* equals the amplitude at some precise moment, or instant, in time. In an AC wave, the instantaneous amplitude varies as the wave cycle progresses. We represent instantaneous amplitude values as points on the wave curves.

The *positive peak* (pk+) *amplitude* equals the maximum value that the instantaneous amplitude attains in the positive direction. The *negative peak* (pk–)

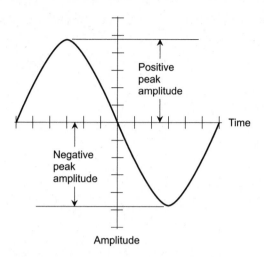

FIGURE 2-9 · Positive and negative peak amplitudes. In this case they are equal.

amplitude equals the maximum value that the instantaneous amplitude attains
in the negative direction.

In some AC waves, the positive and negative peak amplitudes are the same,
but occasionally they differ. Figure 2-9 shows a wave in which the positive peak
amplitude equals the negative peak amplitude. Figure 2-10 illustrates a wave
that has different positive and negative peak amplitudes.

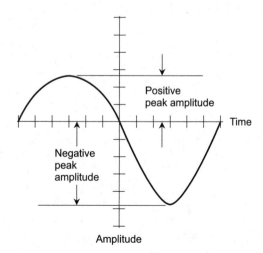

FIGURE 2-10 · A wave in which the positive and
negative peak amplitudes differ.

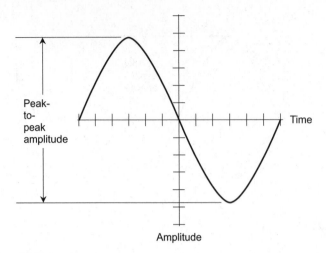

FIGURE 2-11 · Peak-to-peak amplitude

The *peak-to-peak* (pk-pk) *amplitude* of a wave equals the positive peak amplitude minus the negative peak amplitude, which always equals the "amplitude distance" between the positive and negative peaks, as shown in Fig. 2-11. Calculation of peak-to-peak values can get tricky, because we must sometimes subtract a negative number from some other number.

Suppose that each vertical division in Fig. 2-10 represents 1 V. If we scrutinize this graph, we can see that the positive peak voltage (E_{pk+}) equals approximately +2.7 V and the negative peak voltage (E_{pk-}) equals approximately −4.8 V. Therefore, the wave has a peak-to-peak voltage (E_{pk-pk}) of

$$E_{pk\text{-}pk} = E_{pk+} - E_{pk-}$$
$$= +2.7 - (-4.8)$$
$$= 2.7 + 4.8$$
$$= 7.5 \text{ V pk-pk}$$

We can express all three voltages symbolically as follows:

$$E_{pk+} = +2.7 \text{ V pk+}$$
$$E_{pk-} = -4.8 \text{ V pk-}$$
$$E_{pk\text{-}pk} = 7.5 \text{ V pk-pk}$$

In the situation of Fig. 2-11, suppose that we allow each vertical division to represent 5 V. The approximate peak values appear to be

$$E_{pk+} = +24 \text{ V pk+}$$

and

$$E_{pk-} = -24 \text{ V pk}-$$

so therefore

$$E_{pk-pk} = E_{pk+} - E_{pk-}$$
$$= +24 - (-24)$$
$$= 24 + 24$$
$$= 48 \text{ V pk-pk}$$

TIP   *Peak-to-peak amplitude values are never mathematically negative, so we don't have to include plus or minus polarity signs with them. In fact, we shouldn't use a polarity sign when we express a peak-to-peak figure.*

Hybrid AC/DC Waves

In the "real world," an AC wave can have a *DC component* superimposed. If the *absolute value* of the DC component exceeds the absolute value of the positive or negative peak amplitude (whichever is larger) of the AC wave, then we get *pulsating DC* or *fluctuating DC*. We define the absolute value of an electrical quantity as mathematicians define the absolute value of a variable. The absolute value of x, symbolized $|x|$, represents the "distance of x from zero." Therefore

$$\text{If } x \geq 0, \text{ then } |x| = x$$

and

$$\text{If } x < 0, \text{ then } |x| = -x$$

Imagine a 200-V DC source connected in series with an AC source that has approximate peak voltages of $E_{pk+} = +165$ V and $E_{pk-} = -165$V. Pulsating DC appears at the output with an average value of 200 V, but with instantaneous values much higher and lower than 200 V. Figure 2-12 illustrates the waveform in this situation. We portray the DC component as a straight, horizontal, dashed line, and the AC component as a sine curve.

Root-Mean-Square Values

Often, we'll want to express the *effective amplitude* of an AC wave instead of the peak or peak-to-peak amplitudes. Engineers define the effective amplitude of an AC wave as the voltage, current, or power that a DC source would have to generate in order to produce the same amount of heating in a plain *resistive load* (such as an electric space heater) as the AC wave does.

The most common specification for effective amplitude goes by the jargon *root-mean-square* (RMS). This term comes from the mathematical operations

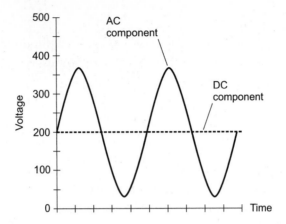

FIGURE 2-12 · A composite wave resulting from DC superimposed on AC.

that we perform on the instantaneous values over a period of time equivalent to many cycles. First we square all the instantaneous amplitude values (infinitely many such points exist in theory, but we can get a good result by taking only a few million!), then we average them all, and finally we take the square root of the result. That's a lot of arithmetic, but with most AC or pulsating DC waveforms, a computer can do the work for us.

A perfect sine wave without any DC component has an RMS amplitude equal to about 0.7071 times the peak amplitude, or 0.3536 times the peak-to-peak amplitude. For a perfect square wave with no DC component, the RMS amplitude equals the positive peak amplitude. For a sawtooth or irregular wave with no DC component, the relationship between the RMS and peak amplitudes depends on the shape of the waveform (the *waveform function*); we must express this function in mathematical terms if we want to accurately calculate the RMS amplitude.

Still Struggling

The existence of superimposed DC on a wave affects its RMS amplitude. The task of figuring out the RMS value in such situations can get complicated, but nothing that a good computer program can't handle! In any event, the RMS amplitude can never exceed the absolute value of the positive or negative peak amplitudes, whichever is larger.

PROBLEM 2-3

Suppose that a sine wave has a peak-to-peak amplitude of 200 V, and consists of a pure AC wave with +50 V DC superimposed. What's the positive peak voltage? What's the negative peak voltage?

SOLUTION

The peak voltages are +100 V and −100 V with respect to the DC component. Because the DC component equals +50 V, the positive and negative peak voltages work out as

$$E_{pk+} = (+50\ V) + (+100\ V)$$
$$= +150\ V\ pk+$$

and

$$E_{pk-} = (+50\ V) + (-100V)$$
$$= -50\ V\ pk-$$

PROBLEM 2-4

How much positive DC voltage must we superimpose on an AC wave with E_{pk+} = +150 V and E_{pk-} = −50 V to obtain a fluctuating DC voltage as opposed to an alternating voltage (that is, to prevent the polarity from ever reversing)? How much negative voltage must we superimpose to obtain a negative fluctuating DC voltage instead of an alternating voltage?

SOLUTION

We must superimpose +50 V DC or −150 V DC to prevent the polarity from reversing. The DC voltage of +50 V brings the negative peak voltage up to zero, and the DC voltage of −150 V brings the positive peak voltage down to zero. In either of these situations, the polarity never reverses, so we have fluctuating DC voltages.

Phase Relationships

When two sine waves have the same frequency, they can behave differently if their cycles begin at different times. Whether or not the *phase difference* (also called the *phase angle* and usually specified in degrees) has any significance depends on the nature of the circuit.

Waves in Phase Coincidence

The phase relationship between two waves has meaning only when the frequencies are identical. If the frequencies differ, the relative phase constantly changes, so we can't obtain a definite value for it. In the following discussions of phase angle, let's assume that the two waves always have the same frequency.

The term *phase coincidence* means that two waves having identical frequencies begin at exactly the same instant in time, so their waveforms follow each other along from instant to instant. Figure 2-13 shows an example of phase coincidence between two waves whose amplitudes differ. The phase difference in this case equals 0°.

If two perfect sine waves exist in phase coincidence, the positive peak amplitude of the resultant wave, which also happens to be a sine wave, equals the sum of the positive peak amplitudes of the two composite waves. Also, the negative peak amplitude of the resultant wave equals the sum of the negative peak amplitudes of the composites, and the peak-to-peak amplitude of the resultant wave equals the sum of the peak-to-peak amplitudes of the composites. The phase of the resultant wave coincides with the phases of the two composites.

Waves 180° Out of Phase

When two pure sine waves of identical frequency begin exactly 1/2 cycle apart in time, we say that they occur *180° out of phase* with respect to each other. Figure 2-14 illustrates a situation of this sort.

If two sine waves have the same amplitude and exist 180° out of phase, and if neither of them has a DC component, then they completely cancel each other

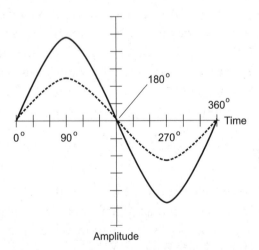

FIGURE 2-13 · Two sine waves that occur in phase coincidence.

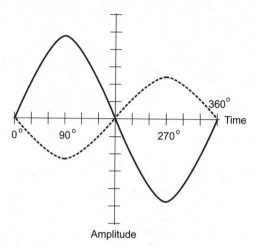

FIGURE 2-14 · Two sine waves that occur 180° out of phase.

out because the instantaneous amplitudes of the two waves are equal and opposite at every moment in time. If the two sine waves have different amplitudes, neither of them has a DC component, and they occur 180° out of phase, then the peak-to-peak amplitude of the resultant equals the difference between the peak-to-peak amplitudes of the two composite waves, and the phase of the resultant wave coincides with the phase of the stronger of the two composites.

A sine wave has the unique property that, if we shift its phase by 180°, we'll obtain the same result as we get if we *invert* the original wave (that is, if we "turn it upside-down"). A perfect, symmetrical square wave also has this property. However, most other waveforms don't behave this way. We cannot, in general, say that a 180° phase shift is equivalent to *phase opposition* (inverting the waveform).

Still Struggling

Theoretically, moving a wave forward or backward by 1/2 cycle doesn't change it in the same way as inverting it does. A pure sine wave with no DC component, or a symmetrical square wave with no DC component, constitute special cases where the two actions give us the same practical result in ordinary electrical circuits. However, with other waves, the results usually differ. To see these effects, graph some examples of non-sine waves or waves with DC components, and then compare the results of a 180° phase shift with outright wave inversion.

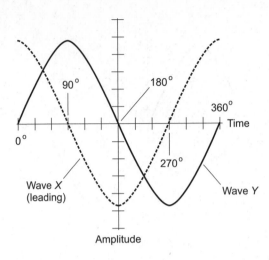

FIGURE 2-15 • Wave *X* leads wave *Y* by 90°.

Leading Phase

Two waves can differ in phase by any amount from 0° (phase coincidence) through 180° to 360° (phase coincidence again, as long as the signal continues indefinitely). Consider two sine waves, *X* and *Y*, with identical frequencies. If wave *X* begins a fraction of a cycle earlier than wave *Y*, then we say that wave *X leads* wave *Y* in phase, or that wave *X* has the *leading phase*. For this definition to work, *X* must begin its cycle less than 180° before *Y*. Figure 2-15 shows wave *X* leading wave *Y* by 90°.

TIP *If wave X (the dashed curve in Fig. 2-15) leads wave Y (the solid curve), then wave X lies somewhat to the left of wave Y in a graphical display. In a horizontal time-line display, the left-hand direction represents the past (earlier times), the origin or center point represents the present moment in time, and the right-hand direction represents the future (later times).*

Lagging Phase

Suppose that a wave *X* begins its cycle more than 180°, but less than 360°, ahead of another wave *Y*. In this situation, we can imagine that wave *X* starts its cycle later than wave *Y*, by some length of time corresponding to a phase angle between 0° and 180°. In that case, we say that wave *X lags* wave *Y* in phase, or that wave *X* has the *lagging phase*. Figure 2-16 shows wave *X* lagging wave *Y* by 90°.

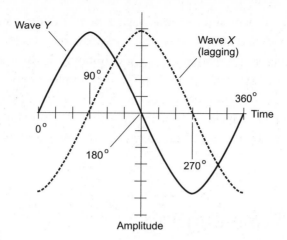

FIGURE 2-16 · Wave *X* lags wave *Y* by 90º.

If two waves, both continuing for an indefinite length of time, have the same frequency and different phase, how do we know that one wave leads the other by part of a cycle, instead of lagging by a cycle and a fraction, or by a few hundred, thousand, or million cycles and a fraction? The answer lies in the real-life effects. By convention, engineers always express phase differences as values between 0° and 180°, either lagging or leading. Sometimes this range is given as −180° to +180° (lagging to leading).

TIP *Because wave* X *(the dashed curve in Fig. 2-16) lags wave* Y *(the solid curve), wave* X *lies somewhat to the right of wave* Y *on the graph. As we move toward the left on any such graph, we go earlier in time; as we move toward the right, we go later in time. Imagine time as "flowing" from left to right. Don't let graphs of this sort confuse you!*

Vector Diagrams

In mathematics, physics, and engineering, *vectors* represent quantities that have two independent characteristics: *magnitude* and *direction*. We can draw vectors as arrowed line segments in diagrams or graphs, with the length denoting the magnitude and the orientation representing the direction. In text, we denote vectors in boldface. Vector diagrams often prove useful in illustrating phase relationships between sine waves that have the same frequency but not necessarily the same amplitude.

If a sine wave X leads another sine wave Y by $q°$ (where q represents some number between 0 and 180), then we can draw the two waves as vectors, with vector X pointing in a direction $q°$ of arc counterclockwise from vector Y. If wave X lags Y by $q°$ (where q, again, equals some number between 0 and 180), then vector X points in a direction clockwise from vector Y by $q°$ of arc. If two waves exist in phase coincidence, their vectors overlap (line up). If they occur 180° out of phase, they point in exactly opposite directions.

Still Struggling

The graphs of Fig. 2-17 show four phase relationships between two hypothetical waves X and Y. Wave X always has twice the amplitude of wave Y, so vector X has twice the length of vector Y. At A, wave X is in phase coincidence with wave Y. At B, wave X leads wave Y by 90°. At C, waves X and Y occur 180° out of phase. At D, wave X lags wave Y by 90°. As time passes in each of these four scenarios, both vectors rotate counterclockwise at identical, constant angular speed, undergoing one full 360° turn with every complete wave cycle.

PROBLEM **2-5**

Suppose that two sine waves called X and Y have identical frequencies and follow each other along in phase coincidence. Both waves have 20 V pk-pk with no DC component. What are the positive peak, negative peak, and peak-to-peak voltages of the composite wave?

SOLUTION

The composite wave has twice the peak-to-peak amplitude of either wave alone, that is, $20 \times 2 = 40$ V pk-pk. Because neither wave has a DC component, both waves exhibit +10 V pk+ and −10 V pk−. The positive peak amplitude of the composite equals twice the positive peak amplitude of either wave alone, that is, $+10 \times 2 = +20$ V pk+. The negative peak amplitude of the composite equals twice the negative peak amplitude of either wave alone, that is, $−10 \times 2 = −20$ V pk−.

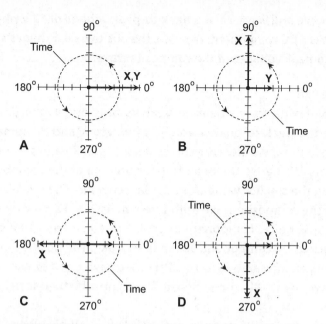

FIGURE 2-17 · At A, waves *X* and *Y* are in phase, so vectors **X** and **Y** point in the same direction. At B, wave *X* leads wave *Y* by 90°, so vector **X** runs 90° counterclockwise from vector **Y**. At C, waves *X* and *Y* occur 180° out of phase, so vectors **X** and **Y** point in opposite directions. At D, wave *X* lags wave *Y* by 90°, so vector **X** runs 90° clockwise from vector **Y**.

PROBLEM 2-6

Suppose that two sine waves called *X* and *Y* have identical frequencies and occur 180° out of phase. Both waves have peak-to-peak voltages of 20 V pk-pk, with no DC component. What are the positive peak, negative peak, and peak-to-peak voltages of the composite wave?

SOLUTION

The two waves completely cancel each other out. Therefore, the positive peak, negative peak, and peak-to-peak voltages all equal zero.

PROBLEM 2-7

Suppose that two sine waves called *X* and *Y* have identical frequencies and occur 180° out of phase. Wave *X* has a peak-to-peak voltage of

20 V pk-pk, and wave *Y* has a peak-to-peak voltage of 12 V pk-pk. Neither wave has a DC component. What are the positive peak, negative peak, and peak-to-peak voltages of the composite wave?

✔ **SOLUTION**

The composite wave's peak-to-peak voltage equals the difference between the peak-to-peak voltages of waves *X* and *Y*. Remember that by convention, we always express peak-to-peak values as nonnegative numbers. We derive the peak-to-peak voltage of the composite by subtracting the smaller value from the larger value. The peak-to-peak voltage of the composite wave therefore equals 20 − 12 = 8 V pk-pk.

Because no DC component exists in wave *X*, it has +10 V pk+. Wave *Y* also lacks a DC component, so it has −6 V pk−. The positive peak of wave *X* occurs at the same instant as the negative peak of wave *Y* because the waves exist 180° out of phase. The composite therefore has +10 − 6 = +4 V pk+.

Wave *X* has no DC component, so it has −10 V pk−. Wave *Y* also lacks a DC component, so it has +6 V pk+. The negative peak of wave *X* occurs at the same instant as the positive peak of wave *Y*. Because the waves exist 180° out of phase, the composite has −10 + 6 = −4 V pk−.

Utility Power Transmission

Electrical energy undergoes several transformations as it gets from the point of origin, usually an *electric generator*, to the end users. The initial source comprises potential or kinetic energy in some nonelectrical form, such as falling or flowing water (*hydroelectric energy*), coal or oil (*fossil-fuel energy*), radioactive substances (*nuclear energy*), moving air (*wind energy*), light or heat from the sun (*solar energy*), or heat from the earth's interior (*geothermal energy*).

Generating Plants

In fossil-fuel, nuclear, geothermal, and some solar energy generating systems, heat boils water to form pressurized steam that drives *turbines*. The turbines produce the rotational force (*torque*) necessary to drive large *electric generators*. As the power demand rises, it takes increasing mechanical torque to turn the generator shaft at the required speed.

A *photovoltaic (PV) energy-generating system* converts sunlight directly into DC electricity by means of components called *photovoltaic cells*. In order to drive most conventional electrical appliances, we must convert the DC to AC. A *power inverter* can perform the conversion. We need electrochemical batteries to store the DC energy if we want a self-contained PV system to provide useful power at night or on dark days. Photovoltaic systems primarily serve homes and small businesses. Self-contained PV systems (also called *stand-alone PV systems* or *off-grid PV systems*) produce no waste products other than chemicals that must be discarded when storage batteries wear out. A PV system can operate in conjunction with existing utilities, without storage batteries, to supplement the total available energy supply to all consumers in a power grid. We call this arrangement an *interactive PV system* or *a grid-intertie PV system*. It causes no environmental pollution during operation.

In a *hydroelectric power plant*, the movement of water (waterfalls, tides, river currents, or ocean currents) drives turbines that turn the generator shafts. In a *wind-driven electric power plant*, moving air operates *wind turbines* similar to windmills. These power plants do not pollute directly. However, large dams can disrupt ecosystems, have adverse effects on agricultural and economic interests downstream, and displace people upriver by flooding their land. Some people regard large wind-turbine arrays as eyesores. They can also present a threat to migrating birds.

TIP *Whatever source we use to generate the electricity at a large power plant, we get AC output on the order of several hundred kilovolts, and in some cases a megavolt or more.*

High-Tension Lines

When electric power travels over wires for long distances, *ohmic resistance* in the conductors dissipates some of that power as heat. Engineers attempt to minimize this loss in two ways. First, they keep the wire resistance to a minimum by using large-diameter wires made from metal having excellent conductivity, and by routing power lines to keep them as short as possible. Second, engineers make the transmission-line voltage as high as possible. Some people call long-distance, high-voltage power lines *high-tension lines*.

We can understand why high voltage minimizes power-transmission line loss when we scrutinize the equations relating power, current, voltage, and resistance. We express power loss in terms of voltage and resistance as

$$P_{\text{loss}} = I^2 R$$

where P_{loss} represents the power (in watts) dissipated in the line as heat, I represents the line current (in amperes), and R represents the line resistance (in ohms). For a given span of electrical transmission line, let's assume that the value of R remains constant. The value of P_{loss} therefore varies in proportion to the square of the current in the line, which depends on the *net load*—the total power that all the end users demand as a group.

Power-transmission line current also depends on the line voltage. Current varies inversely in proportion to voltage at a given fixed power load. The formula is

$$I = P_{load}/E$$

where I represents the line current (in amperes), P_{load} represents the total power demanded by the end users (in watts), and E represents the line voltage (in volts). When we substitute from this equation into the previous one, we obtain

$$P_{loss} = I^2 R$$
$$= (P_{load}/E)^2 R$$
$$= (P_{load})^2 R/E^2$$

For any given fixed values of P_{load} and R, doubling E reduces P_{loss} to one-quarter (25%) of its previous value. If we multiply the line voltage by a factor of 10, the theoretical line power loss drops to 1/100 (1%) of its former value. Obviously, it makes sense to generate the highest possible voltage if we want to have efficient long-distance electric power transmission. As we increase the voltage, we "get our reward squared"!

We can reduce the loss in a power line even further by using DC, rather than AC, for long-distance transmission. Direct currents don't produce *electromagnetic (EM) fields* as alternating currents do, so a secondary source of power-line inefficiency, known as *EM radiation loss*, is eliminated by the use of DC.

Still Struggling

Long-distance DC power transmission has proven difficult and costly to implement. It necessitates high-power, high-voltage rectifiers at a generating plant and DC-to-AC power inverters at distribution stations, where high-tension lines branch out into lower-voltage, local lines.

Transformers

In the United States, most households employ AC voltages of 117 V and 234 V RMS. These comparatively low voltages are obtained from higher voltages by using *step-down transformers* that reduce the voltage of high-tension lines (100 kV or more) down to a few thousand volts for distribution within municipalities. These transformers have considerable bulk and mass because they must carry significant power. Several of them can be placed in a building or a fenced-off area. Their outputs go to power lines that run along city streets.

Smaller transformers, usually mounted on utility poles or underground, step the municipal voltage down to 234 V RMS for distribution to homes and businesses. The 234 V electricity occurs as three separate AC waves, called *phases*, which appear at the distribution box in every building. Each wave runs 120° out of phase with the other two, yielding so-called *three-phase AC*. Large appliances, such as electric stoves, ovens, and laundry machines, operate from three-phase AC electricity at 234 V RMS. Conventional wall outlets in the United States usually carry single-phase AC electricity at 117 V RMS.

PROBLEM 2-8

Suppose that a span of high-tension utility line has an effective resistance of 100 ohms, and the line carries 500 kV RMS AC. What's the power loss?

✔ SOLUTION

We can't answer this question unless we know the total amount of power that the end users demand, or else the current carried by the line. If no power demand exists at the load, then no current flows in the line, so no loss occurs. However, if a substantial power demand exists at the load end of the line, then the line must carry considerable current. As the power demand (and consequently the line current) increases, so does the line loss.

PROBLEM 2-9

Suppose that in the situation described in Problem 2-8, the consumers at the end of the line collectively demand 500 kilowatts (500 kW or 500,000 W). What's the line loss?

SOLUTION

We can use the formula for the power loss P_{loss} in terms of the load power P_{load}, the line resistance R, and the line voltage E. In this instance, we have

$$P_{load} = 500,000 \text{ W}$$
$$R = 100 \text{ ohms}$$
$$E = 500,000 \text{ V}$$

We calculate P_{loss} as

$$P_{loss} = (P_{load})^2 \, R/E^2$$
$$P_{loss} = 500,000^2 \times 100/500,000^2$$
$$= 100 \text{ W}$$

This power line has amazing efficiency, doesn't it? In order to deliver *half a million watts* to customers, the power generator must produce only *100 extra watts* to make up for the ohmic loss in the wires.

TIP   *As you might guess, I chose the numbers for dramatic effect in Problem 2-9 and its solution. In the "real world" of utility power transmission, things rarely work out that well.*

QUIZ

Refer to the text in this chapter if necessary. A good score is eight correct. The answers are listed in the back of the book.

1. Figure 2-18 illustrates a time-domain graph of an AC waveform. Suppose that each horizontal division represents exactly 1 μs, and each vertical division represents exactly 10 V. What's the period of the wave as shown?

    A. 1.5 μs
    B. 3.0 μs
    C. 6.0 μs
    D. We can't say because it's ambiguous.

2. What's the frequency of the wave shown in Fig. 2-18 and described in Question 1, rounded to three significant digits?

    A. 167 kHz
    B. 333 kHz
    C. 667 kHz
    D. We can't say because it's ambiguous.

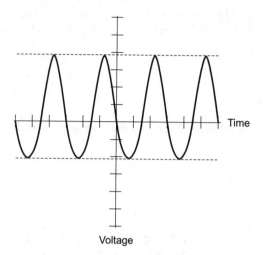

Time

Voltage

FIGURE 2-18 · Illustration for Quiz Questions
1 through 5.

3. **What's the approximate positive peak voltage of the wave shown in Fig. 2-18 and described in Question 1?**
   A. +18 V pk+
   B. +21 V pk+
   C. +39 V pk+
   D. +60 V pk+

4. **What's the approximate negative peak voltage of the wave shown in Fig. 2-18 and described in Question 1?**
   A. −18 V pk−
   B. −21 V pk−
   C. −39 V pk−
   D. −60 V pk−

5. **What's the approximate peak-to-peak voltage of the wave shown in Fig. 2-18 and described in Question 1?**
   A. 18 V pk-pk
   B. 21 V pk-pk
   C. 39 V pk-pk
   D. 60 V pk-pk

6. **What's the peak-to-peak voltage of a pure AC sine wave with no DC component if we measure it as 24.00 V RMS?**
   A. 67.87 V pk-pk
   B. 33.94 V pk-pk
   C. 16.97 V pk-pk
   D. 8.486 V pk-pk

7. What's the RMS voltage of a pure AC sine wave if we measure its positive peak voltage as +24.00 V pk+ and its negative peak voltage as −24.00 V pk−?

   A. 67.87 V RMS
   B. 33.94 V RMS
   C. 16.97 V RMS
   D. 8.486 V RMS

8. What's the angular frequency of a 60.0-Hz sine wave in radians per second?

   A. 9.55 rad/s
   B. 19.1 rad/s
   C. 188 rad/s
   D. 377 rad/s

9. Suppose that a high-tension power line carries 300 kV and has a loss resistance of 10 ohms. The line carries 100 A of current. How much power does the line dissipate as heat, which can never reach the end users?

   A. 10 kW
   B. 30 kW
   C. 100 kW
   D. 3000 kW

10. Imagine that we double the voltage in the high-tension power line described in Question 9 to 600 kV, so that the current demand decreases to half its former value. Assuming that the line resistance does not change, how much power does the line dissipate as heat loss now?

   A. 1500 kW
   B. 750 kW
   C. 50 kW
   D. 25 kW

chapter 3

Impedance and Admittance

Impedance quantifies the extent to which a component or circuit opposes (or *impedes*) the flow of AC. Theoretically, impedance consists of two independent components called *resistance* and *reactance*. Conversely, *admittance* quantifies the extent to which a component or circuit allows (or *admits*) the flow of AC, comprising two independent components known as *conductance* and *susceptance*.

CHAPTER OBJECTIVES

In this chapter you will

- Define the positive square root of −1 as the *j* operator.
- Perform calculations with complex numbers and vectors.
- Discover reactance and see how it, along with resistance, affects AC.
- See how transmission lines affect radio signals.
- Discover susceptance and see how it, along with conductance, affects AC.
- Learn how complex impedances and admittances combine.

The *j* Operator

Imaginary numbers originally got their name because, when people first began to seriously think about them, they seemed like artifacts of a few wild mathematicians' imaginations. However, imaginary numbers are as real as (or, perhaps better stated, no less abstract than) the so-called *real numbers*, such as 2, 5/4, –87, –73.55, or the square root of 10.

Imaginary Numbers

The set of imaginary numbers derives from the square root of –1, called the *unit imaginary number i* by mathematicians and the *j operator* by engineers. It's the number that, when multiplied by itself, results in a product of –1. Let's use the engineering notation and call it *j*. Therefore, we can say that

$$j \times j = -1$$

It also happens that –*j*, when multiplied by itself, gives us –1 because the product of two negatives always produces a positive. We have

$$(-j) \times (-j) = -1$$

The positive square root of –1 differs from the negative square root of –1, even though, when we square either of them, we get –1. (We often forget that 1 has two square roots as well: –1 and 1! In fact, all positive real numbers have two square roots, one of them positive and the other negative.) We should also note that

$$(-j) \times j = 1$$

and

$$j \times (-j) = 1$$

We can multiply *j* by any real number *X*, getting an imaginary number *jX*. Then we can graphically portray the set of all possible imaginary numbers on an *imaginary number line* that resembles the familiar *real number line*. When we place the imaginary number line at a right angle to the real number line and join the two lines at their zero points, we get a coordinate plane as shown in Fig. 3-1.

Complex Numbers

When we add a real number *R* to an imaginary number *jX*, we get a *complex number*. (A better term might be *composite number*, but the people who first worked with these numbers called them complex, and no one has ever changed that term.)

Real numbers are one-dimensional, expressible on a simple straight-line scale, and are therefore called *scalar quantities*. Imaginary numbers are also one-dimensional and can be represented on a simple scale, so they're scalar as well.

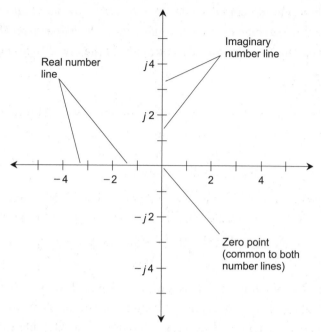

FIGURE 3-1 · We can combine the real number line and the imaginary number line to obtain a two-dimensional coordinate plane.

However, if we want to fully define any complex number, we need two dimensions. For that reason, mathematicians and engineers often portray complex numbers as *vector quantities*. We can express any complex number in the form $R + jX$, where R and X both represent real numbers and j represents the positive square root of -1. We can also portray any complex number as a point on a two-dimensional coordinate plane.

Adding complex numbers requires adding the real parts and the complex parts separately. For example,

$$(4 + j7) + (45 - j83) = (4 + 45) + j(7 - 83)$$
$$= 49 + j(-76)$$
$$= 49 - j76$$

Subtracting complex numbers works in a similar manner, although we mustn't let the signs confuse us! For example,

$$(4 + j7) - (45 - j83) = (4 + j7) + [-(45 - j83)]$$
$$= (4 + j7) + [-1(45 - j83)]$$
$$= (4 + j7) + (-45 + j83)$$
$$= -41 + j90$$

TIP *You can always do complex-number subtraction by negating the second quantity and then adding the two quantities, as we did in the second line of the above calculation. This trick can help you minimize the risk of "sign confusion."*

Complex Vectors

Any complex number $R + jX$ can be represented as a vector in a *complex-number plane*, such as the one we constructed in Fig. 3-1 on page 67. This rendition gives each complex number a unique *magnitude* and a unique *direction*, as shown in Fig. 3-2 and described as follows:

- The magnitude is the distance of the point (R,jX), representing the number $R + jX$, from the origin $(0,j0)$, representing the number $0 + j0$.
- The direction is the angle the vector subtends, measured counterclockwise from the $+R$ axis.

We can define the *absolute value* of a complex number $R + jX$, denoted $|R + jX|$, as the length of its vector (R,jX) in the complex-number plane, measured from the origin $(0,j0)$ to the point (R,jX). We can note the following special cases:

- For a pure positive real number or 0, the absolute value equals the number itself.

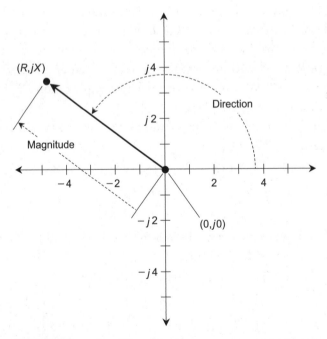

FIGURE 3-2 · Magnitude and direction of a vector in the complex-number plane.

- For a negative real number, the absolute value equals the number multiplied by −1.
- For a pure imaginary number $0 + jX$, the absolute value equals X when X is positive ($X > 0$) and $-X$ when X is negative ($X < 0$).

If a particular complex number is neither pure real nor pure imaginary, then we can find its absolute value with the *Pythagorean theorem* from plane geometry: the formula that we use to find the length of the longest side (or *hypotenuse*) of a right triangle. In this context, we carry out the following steps in order:

- Square R
- Square X (not jX)
- Add the resulting squares
- Take the square root of the sum

This process gives us Z, the length of the vector (R, jX) and therefore its absolute value as shown in Fig. 3-3. Mathematically, we have

$$Z = |R + jX|$$
$$= (R^2 + X^2)^{1/2}$$

where the 1/2 power represents the positive square root.

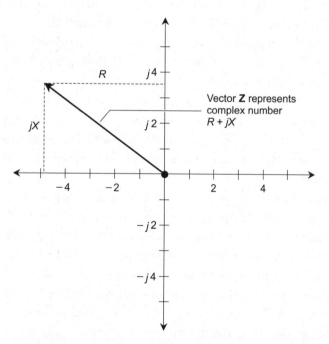

FIGURE 3-3 • The absolute value of a complex number equals the length of its vector in the complex-number plane.

Inductive Reactance

In any DC circuit, resistance constitutes a scalar quantity. Given a constant DC voltage, the current decreases as the resistance increases, in accordance with Ohm's Law. The same law holds for AC in a pure resistance. However, in a component that has *inductance* (such as a coil of wire) or *capacitance* (such as a pair of large metal plates placed close together but not touching), the AC situation gets complicated. Let's look at the situation for inductance now; we'll see what happens with capacitance later in this chapter.

Inductors and Current

If we wind a length of wire into a coil (thereby making it into an *inductor*) and connect it to a source of DC, the coiled-up wire draws an electrical current I that equals the ratio E/R, where I represents the current in amperes, E represents the DC voltage in volts between the ends of the wire, and R represents the DC resistance of the wire in ohms. The coil gets hot as energy dissipates in the wire. If we increase the voltage of the power supply, the wire gets hotter. Eventually, if we leave the wire connected to a substantial power supply for a long enough time, the wire might melt from the heat.

Suppose that we change the voltage across the coil from DC to AC. We start at a low AC frequency, say a few hertz (Hz), and gradually increase the frequency to many kilohertz (kHz), then many megahertz (MHz), and finally into the gigahertz (GHz) range. At every frequency during this process, the coil exhibits a certain *inductive reactance* (denoted X_L), which shows up as "electrical sluggishness." Because of this "sluggishness," it takes some time for the current to establish itself in the coil as each voltage cycle occurs. Engineers express AC inductive reactance in ohms, just as they do with DC resistance. The standard unit of inductance is called the *henry*, abbreviated as a nonitalicized, uppercase letter H.

As we keep increasing the AC frequency in the above-described experiment, we'll reach a point beyond which the current can't get well established in the coil before the polarity of the voltage reverses. As we raise the frequency further still, this effect will become more and more pronounced. Eventually, if the frequency becomes high enough, the coil won't come anywhere near carrying a significant current with each voltage cycle. Therefore, almost no current will flow through the coil. The coil will remain cool even if the supply can deliver so much power that it would melt the coil if it were a DC source.

TIP   *The reactance of an inductor can vary from zero to many megohms. Like pure resistance, inductive reactance affects the current in an AC circuit. But unlike pure resistance, inductive reactance changes with frequency.*

How X_L Varies

If we represent the frequency of an AC source (in hertz) as f, and we represent the inductance of a coil (in henrys) as L, then we can calculate the inductive reactance X_L (in ohms) using the formula

$$X_L = 2\pi f L$$

where π equals approximately 3.14159. This same formula applies if we specify f in kilohertz and L in *millihenrys* (mH), where 1 mH = 0.001 H. It also applies if we input f in megahertz and L in *microhenrys* (μH), where 1 μH = 0.000001 H. If we hold the inductance constant, the inductive reactance increases with increasing AC frequency. If we hold the AC frequency constant, the inductive reactance increases with increasing inductance.

Still Struggling

The value of X_L varies in direct proportion to f, also in direct proportion to L. Figure 3-4 shows "generic" graphs of these relations.

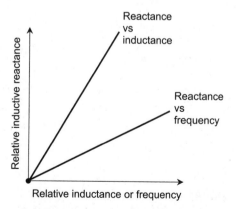

FIGURE 3-4 • Inductive reactance varies in direct proportion to inductance (L) and also to frequency (f).

Points in the RX_L Quarter-Plane

We can plot inductive reactance along a half-line, because it's always non-negative. We can do the same thing with resistance. In a circuit containing both resistance (R) and inductive reactance (X_L), the characteristics become two-dimensional. We can orient the resistance and reactance half-lines perpendicular to each other to define a coordinate grid. In this scheme, RX_L combinations form *complex impedances*. Each point on the grid corresponds to a unique complex impedance value, and each complex impedance value corresponds to a unique point on the grid. We can denote impedances on this so-called RX_L *quarter-plane* in the form $R + jX_L$, where R represents the resistance (always a nonnegative real number) and X_L represents the inductive reactance (another nonnegative real number).

If we have a resistance of, say, $R = 5$ ohms, then the complex impedance equals $5 + j0$, and we can represent it as the point $(5,j0)$ on the RX_L quarter-plane. If we have a pure inductive reactance, such as $X_L = 3$ ohms, then the complex impedance equals $0 + j3$, which lies at the point $(0,j3)$ on the plane. Often, resistance and inductive reactance exist together in the same circuit. In cases like that, we encounter complex impedance values, such as $2 + j3$ or $4 + j1.5$. Figure 3-5 shows some points on the RX_L quarter-plane, each of which represents a specific complex impedance.

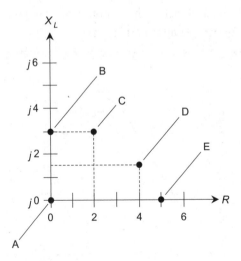

FIGURE 3-5 • Some points in the RX_L imped-ance quarter-plane. At A, $0 + j0$. At B, $0 + j3$. At C, $2 + j3$. At D, $4 + j1.5$. At E, $5 + j0$.

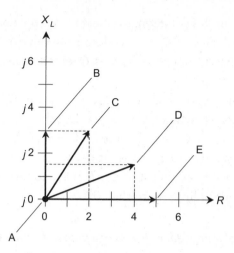

FIGURE 3-6 · Some vectors in the RX_L impedance quarter-plane. At A, $0 + j0$. At B, $0 + j3$. At C, $2 + j3$. At D, $4 + j1.5$. At E, $5 + j0$.

Vectors in the RX_L Quarter-Plane

Engineers represent points in the RX_L quarter-plane as vectors, giving each point a specific and unique magnitude and direction. We can draw lines from the origin out to the points in Fig. 3-5 to obtain geometric rays, each having a certain magnitude and a direction (angle counterclockwise from the R axis). These rays, going out to the points, represent *complex impedance vectors*. Figure 3-6 shows the vectors that we get when we draw rays for the five points of Fig. 3-5.

Current and Voltage in *RL* Circuits

An electronic component containing inductance stores electrical energy as a *magnetic field*. After we connect a source of DC voltage to a coil, it takes awhile for the current to build up to full value. Therefore, if we impose an AC voltage across a coil, the current lags behind the voltage in phase. The extent of the phase lag depends on the AC frequency: The higher the frequency, the greater the phase lag in a constant inductance.

Inductance and Resistance

Imagine that we connect a source of AC voltage across a low-resistance coil, and the AC frequency is so high that the inductive reactance X_L greatly exceeds the DC resistance R. In this situation, the current lags the voltage by almost 90° (1/4 cycle). When the value of X_L is much larger the value of R, the RX_L quarter-plane vector points nearly along the X_L axis, oriented at an angle of almost 90° with respect to the R axis.

When the resistance in a so-called *resistance-inductance (RL) circuit* is significant compared with the reactance, the current lags the voltage by some amount less than 90° (Fig. 3-7). If R is small compared with X_L, the current lag nearly equals 90°; as R gets relatively larger, the lag decreases. When R gets many times larger than X_L, the vector lies almost on the R axis, and the *resistance-inductance (RL) phase angle* is only a little more than 0°.

TIP   *When the complex impedance becomes a pure resistance, the current comes into phase with the voltage, and the vector lies exactly along the R axis. Then we no longer have an RL circuit, but simply an R circuit.*

Calculating Phase Angle in *RL* Circuits

You can use a ruler that has centimeter (cm) and millimeter (mm) markings, along with a protractor, to approximate the phase angle in an *RL* circuit. First, draw a horizontal line a little more than 100 mm long, going from left to right. Construct a vertical line off the left end of this first line, going straight upwards.

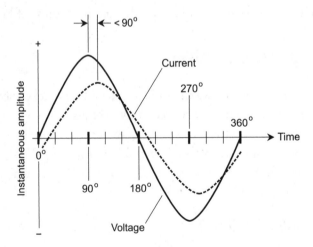

FIGURE 3-7 • In an *RL* circuit, the current lags the voltage by some phase angle greater than 0° but less than 90°.

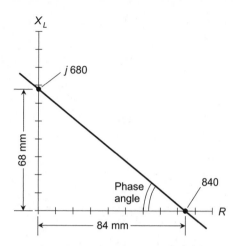

FIGURE 3-8 · Pictorial method of finding the phase angle in an *RL* circuit.

Make this line at least 100 mm long. The horizontal line forms the R axis. The vertical line forms the X_L axis (Fig. 3-8).

If you know the values of X_L and R, divide them down or multiply them up so that they both equal some value between 0 and 100. For example, if $X_L = 680$ ohms and $R = 840$ ohms, divide them both by 10 to get $X_L = 68$ and $R = 84$. Plot these points by making "hash marks" on the axes. In this example, the R mark should lie 84 mm to the right of the origin, and the X_L mark should lie 68 mm up from the origin. Draw a line connecting the marks, as shown in Fig. 3-8. Along with the two axes, this line forms a right triangle. Measure the angle between the slanted line and the R axis. This angle equals the RL phase angle, which we can symbolize as ϕ_{RL}. (The character ϕ is the lowercase Greek letter phi, pronounced "fie" or "fee.")

You can draw the vector corresponding to a complex impedance $R + jX_L$ by constructing a rectangle using the origin and "hash marks" as three of the four vertices, and drawing horizontal and vertical lines to complete the figure. The vector is the diagonal of this rectangle, as shown in Fig. 3-9. The phase angle ϕ_{RL} is the angle between the vector and the R axis.

You can calculate a more exact value for ϕ_{RL}, especially if it happens to be close to 0° or 90° (making pictorial constructions difficult), using a calculator that has inverse trigonometric functions. The RL phase angle equals the *arctangent* (arctan or tan^{-1}) of the ratio of inductive reactance to resistance, as follows:

$$\phi_{RL} = \arctan{(X_L/R)}$$

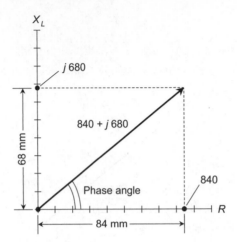

FIGURE 3-9 · Illustration of an RX_L impedance vector.

PROBLEM 3-1

Suppose that a circuit contains 30 ohms of resistance and 40 ohms of inductive reactance. What's the complex impedance? What's the phase angle, rounded off to the nearest degree?

SOLUTION

In this example, $R = 30$ and $X_L = 40$. The complex impedance Z is therefore

$$Z = R + jX_L$$
$$= 30 + j40$$

The phase angle ϕ_{RL} is

$$\phi_{RL} = \arctan(X_L / R)$$
$$= \arctan(40 / 30)$$
$$= \arctan 1.333$$
$$= 53°$$

PROBLEM 3-2

Suppose that the frequency in the above example equals 1000 Hz. What inductance, to the nearest tenth of a millihenry, corresponds to a reactance of 40 ohms in this case?

 SOLUTION

Let's "plug in" the known values to the equation for inductive reactance in terms of frequency and inductance. The general equation is

$$X_L = 2\pi fL$$

When we input the numbers, we get

$$40 = 2 \times 3.14159 \times 1000 \times L$$

We can find L by dividing both sides of the above equation by $2 \times 3.14159 \times 1000$, and then calculating it out as

$$L = (40)/(2 \times 3.14159 \times 1000)$$
$$= 40/6283.18$$
$$= 0.00636620 \text{ H}$$
$$= 6.4 \text{ mH}$$

Capacitive Reactance

Inductive reactance has a "fraternal twin" in *capacitive reactance*, which also constitutes a scalar quantity. Capacitive reactance starts at the same zero point as inductive reactance, but runs off in the opposite direction, having negative values.

Capacitors and Current

Imagine two gigantic, flat, parallel metal plates, both of which conduct electricity perfectly. These plates form a *capacitor*. (Some old texts use the term *condenser* instead.) If supplied with DC, the plates charge up and eventually acquire a potential difference equal to the DC source voltage. The current, once the plates attain the maximum amount of electrical charge, drops to zero.

Now suppose that we change the power supply from DC to AC. Imagine that we can adjust the frequency of this AC from a few hertz to many megahertz. At first, the charges on the plates follow almost exactly along with the applied voltage as the AC polarity reverses; the pair of plates acts like an open circuit. As the frequency increases, the charge can't get well established with each cycle, so current keeps flowing into and out of the plates as the voltage polarity goes back and forth. When the frequency becomes extremely high, the plates don't even come close to getting fully charged with each cycle, so the current flows continuously in and out of them, and the pair of plates behaves almost like a short circuit.

Capacitive reactance (symbolized X_C) quantifies the opposition that a capacitor offers to AC. Like inductive reactance, X_C varies with frequency. But X_C is, by convention, assigned negative ohmic values rather than positive ones. The value of X_C increases negatively as the frequency goes down. The standard unit of capacitance is the *farad*, abbreviated by a nonitalicized, uppercase letter F.

Still Struggling

Sometimes, engineers talk or write about X_C in terms of its absolute value, with the minus sign removed. In complex-number impedance calculations, however, we should always regard X_C as a nonpositive quantity (that is, negative or zero). This convention prevents us from getting X_C confused with inductive reactance X_L that always has a nonnegative value (positive or zero).

How X_C Varies

In some ways, capacitive reactance behaves like a mirror image of inductive reactance. In another sense, we can imagine X_C as an extension, rather than a reflection, of X_L into negative values.

If we specify the frequency of an AC source (in hertz) as f, and we specify the value of a *capacitor* (in farads) as C, then we can calculate the capacitive reactance X_C (in ohms) using the formula

$$X_C = -1/(2\pi fC)$$
$$= -(2\pi fC)^{-1}$$

This formula also applies if we express the frequency in megahertz and the capacitance in *microfarads* (μF), where 1 μF = 0.000001 F. Capacitive reactance varies inversely with the negative of the frequency, and also inversely with the negative of the capacitance.

TIP   *You might wonder, "What happened to millifarads (mF) in this discussion?"*
For some reason, engineers almost never use millifarads to quantify capacitance.

In fact, the appearance of the abbreviation mF can cause confusion. If you ever see a component labeled "100 mF," the engineers probably mean to specify a value of 100 µF—but you'd better check their text references to make sure!

Still Struggling

The function X_C versus f appears as a curve when graphed. This curve approaches a *singularity* (in lay terms, it "blows up") as the frequency nears zero. The function of X_C versus C also appears as a curve that "blows up" as the capacitance approaches zero. Figure 3-10 shows generic graphs of these functions.

Points in the RX_C Quarter-Plane

In a circuit containing resistance and capacitive reactance, the characteristics are two-dimensional, in a way that "mirrors" the situation with the RX_L quarter-plane. We place the resistance ray and the capacitive-reactance ray end-to-end at right angles to make the RX_C *quarter-plane* (Fig. 3-11). We plot resistance horizontally, with increasing values going toward the right. We plot capacitive reactance vertically, with increasingly negative values going downward. Each point corresponds to exactly one complex impedance value. Conversely, each

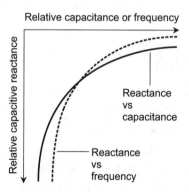

FIGURE 3-10 · Capacitive reactance varies inversely according to the negatives of the frequency (f) and capacitance (C).

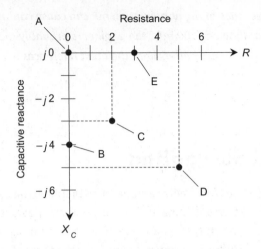

FIGURE 3-11 · Some points in the RX_C impedance quarter-plane. At A, $0 - j0$. At B, $0 - j4$. At C, $2 - j3$. At D, $5 - j5$. At E, $3 - j0$.

complex impedance value coincides with exactly one point on the plane. We render impedance values that contain resistance and capacitance in complex-number form, just as we do with impedances that contain resistance and inductive reactance. Because X_C can never attain positive values, engineers often denote resistance and capacitive reactance combinations in the form $R - jX_C$.

If the resistance is pure, say $R = 3$ ohms, then the complex impedance is $3 - j0$, corresponding to the point $(3, j0)$ on the RX_C quarter-plane. If we have a pure capacitive reactance, say $X_C = -4$ ohms, then the complex impedance is $0 - j4$, and its corresponding point lies at $(0, -j4)$ on the RX_C plane. If nonzero resistance and capacitive reactance both exist, we have complex impedances, such as $2 - j3$ or $5 - j5$.

Vectors in the RX_C Quarter-Plane

Figure 3-11 shows several complex impedance points, corresponding to various combinations of resistance and capacitive reactance. Each point represents a certain distance to the right of the origin $(0, j0)$, along with a certain displacement downward from the origin.

We can render impedance points in the RX_C quarter-plane as vectors, just as we do in the RX_L quarter-plane. The magnitude of a vector equals the distance of its end (terminating) point from the origin; the direction is the angle measured clockwise from the resistance (R) axis, which we specify in *negative degrees*. Figure 3-12 portrays the vectors corresponding to the points of Fig. 3-11.

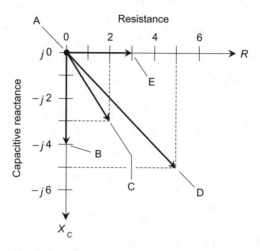

FIGURE 3-12 • Some vectors in the RX_C imped-ance quarter-plane. At A, $0 - j0$. At B, $0 - j4$. At C, $2 - j3$. At D, $5 - j5$. At E, $3 - j0$.

Current and Voltage in *RC* Circuits

A capacitor stores energy in the form of an *electric field*. When we impose an AC voltage across a circuit containing resistance and capacitance, the current leads the voltage in phase. The phase angle always exceeds 0° but remains less than 90°.

Capacitance and Resistance

When the resistance in a *resistance-capacitance (RC) circuit* is significant com-pared with the absolute value of the capacitive reactance, the current leads the voltage by something less than 90°, as shown in Fig. 3-13. If R is small compared with the absolute value of X_C, the difference equals almost 90°. As R gets larger relative to the absolute value of X_C, the phase difference decreases.

The value of R can increase relative to the absolute value of X_C because we deliberately put resistance into a circuit, or because we make the AC frequency so low that the absolute value of X_C rises to a value comparable with the *leakage resistance* of the capacitor. In either case, we can represent a situation of this kind as a resistor and capacitor connected in series.

As the resistance in an *RC* circuit grows until it greatly exceeds the absolute value of the capacitive reactance, the phase angle approaches 0°. The same thing happens if the absolute value of X_C gets small compared with R.

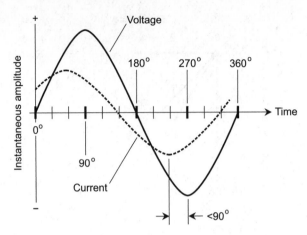

FIGURE 3-13 · In an *RC* circuit, the current leads the voltage by some phase angle greater than 0° but less than 90°.

TIP *When R is many times larger than the absolute value of* X_C *, the vector in the* RX_C *quarter-plane lies almost along the R axis. Then the* RX_C *phase angle is close to 0°. Ultimately, if the absolute value of the capacitive reactance gets vanishingly small, the circuit behaves as a pure resistance, the current comes into phase with the voltage, and the impedance vector lies along the R axis.*

Still Struggling

We constantly refer to the absolute value of X_C in this context, and not X_C itself, because capacitive reactance is always negative. If we take its absolute value, we get positive quantities that we can meaningfully compare with resistance values, which are always positive.

Calculating Phase Angle in *RC* Circuits

If you know the ratio X_C/R in an *RC* circuit, then you can find the *resistance-capacitance (RC) phase angle* using a protractor and a ruler, just as you can do with *RL* circuits, as long as the angles aren't too close to 0° or 90°.

Draw a line somewhat longer than 100 mm, going from left to right on the paper. Then use the protractor to construct a line going somewhat more than

100 mm vertically downwards, starting at the left end of the horizontal line. The horizontal line forms the R axis of the RX_C quarter-plane. The line going down forms the X_C axis. Figure 3-14 illustrates an example.

If you know the actual values of X_C and R, divide or multiply them by a constant, chosen to make both values fall between −100 and 100. For example, if $X_C = -3800$ ohms and $R = 7400$ ohms, divide them both by 100, getting −38 and 74. Plot these points on the lines as "hash marks." In the scenario of Fig. 3-14, the X_C mark goes 38 mm down from the origin, while the R mark goes 74 mm to the right of the origin. Draw a slanted line connecting the marks. Measure the angle between the slanted line and the R axis. This angle will lie somewhere between 0° and 90°. Multiply it by −1 to get the phase angle for the RC circuit, symbolized ϕ_{RC}.

You can diagram the complex RX_C impedance vector by constructing a rectangle using the origin and the "hash marks," drawing new perpendicular lines to complete the figure. The diagonal of this rectangle, running out from the origin (Fig. 3-15), represents the vector. The phase angle is the negative of the angle between the R axis and the vector.

The more accurate way to find RC phase angles involves trigonometry. Determine the ratio X_C/R and enter it into a calculator. This ratio should yield a negative number or zero, because X_C is always negative or zero and R is always positive. Find the arctangent (arctan or $\tan^{-1}$) of this number to get ϕ_{RC}. Mathematically, the situation works out according to the formula

$$\phi_{RC} = \arctan (X_C/R)$$

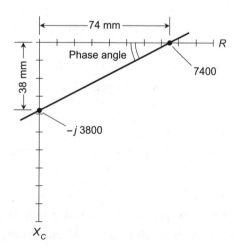

FIGURE 3-14 • Pictorial method of finding the phase angle in an RC circuit.

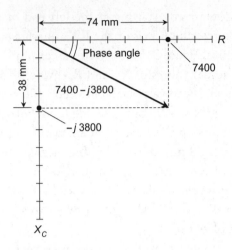

FIGURE 3-15 • Illustration of an RX_C impedance vector.

PROBLEM 3-3

A circuit contains 100 ohms of resistance and −100 ohms of capacitive reactance. What's the complex impedance? What's the phase angle, rounded to the nearest degree?

SOLUTION

In this example, $R = 100$ and $X_C = -100$. The complex impedance Z is therefore

$$Z = 100 - j100$$

The phase angle ϕ_{RC} is

$$\phi_{RC} = \arctan (-100/100)$$
$$= \arctan -1$$
$$= -45°$$

PROBLEM 3-4

Suppose that the frequency in the above example is 1000 kHz. What capacitance, to the nearest ten-thousandth of a microfarad, corresponds to a reactance of −100 ohms?

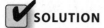SOLUTION

Note that 1000 kHz = 1.000 MHz. First, let's "plug in" the known values for capacitive reactance in terms of frequency (in megahertz) and capacitance (in microfarads). The applicable formula is

$$X_C = -1/(2\pi f C)$$

When we put the values in, we get

$$-100 = -1/(2 \times 3.14159 \times 1.000 \times C)$$

Multiplying each side by −1 gives us

$$100 = 1/(2 \times 3.14159 \times 1.000 \times C)$$

We can invert (take the reciprocal of) both sides of the above equation to obtain

$$0.01 = 2 \times 3.14159 \times 1.000 \times C$$

We can calculate C by dividing both sides of the equation by $2 \times 3.14159 \times 1.000$, and then doing the arithmetic to arrive at the solution

$$C = (0.01)/(2 \times 3.14159 \times 1.000)$$
$$= 0.01/6.28318$$
$$= 0.00159155 \ \mu F$$
$$= 0.0016 \ \mu F$$

The *RX* Half-Plane

The RX_L quarter-plane, as we've defined it here, forms the upper portion of an *RX half-plane* as shown in Fig. 3-16. Similarly, the quarter-plane for R and X_C forms the lower portion of an RX half-plane. In this coordinate grid, we represent resistance values as nonnegative real numbers going off to the right along the horizontal axis. Reactance values, whether inductive (positive) or capacitive (negative), correspond to imaginary numbers, going up and down along the vertical axis. We can therefore portray any specific complex impedance $R + jX$ as a unique complex number, where R can represent any nonnegative real number and X can represent any real number.

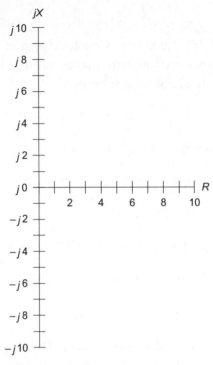

FIGURE 3-16 • The complex impedance
(*RX*) half-plane.

Sometimes the italicized, uppercase letter *Z* appears in place of the word "impedance" in general discussions, and yet the figures given are real numbers, not complex numbers. For example, suppose that you see a dynamic loudspeaker advertised as having "*Z* = 8 ohms." If no specific complex impedance is given, then *Z* can, in theory, correspond to $8 + j0, 0 + j8, 0 - j8$, or any value on a half-circle of points in the *RX* half-plane that lie 8 units away from the origin (Fig. 3-17). When an impedance *Z* is represented as the length (or absolute value) of its vector only, without specifying the reactance or resistance, *Z* is sometimes called an *absolute-value impedance*.

TIP *If an author (or an advertisement) doesn't tell us which complex impedance is meant when quoting a single-number ohmic figure, the number usually refers to a* nonreactive impedance *in which the imaginary, or reactive, factor equals zero. We can also call it a* resistive impedance *or a* pure resistance.

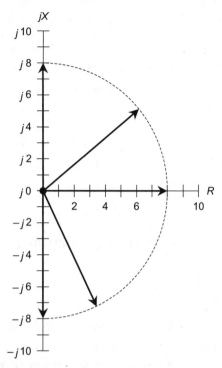

FIGURE 3-17 · Several different vectors in the *RX* half-plane. Theoretically, each vector represents $Z = 8$ ohms.

Impedances in Series

Given two impedances $Z_1 = R_1 + jX_1$ and $Z_2 = R_2 + jX_2$ connected in series, the net complex impedance Z equals their vector sum, given by the equation

$$Z = (R_1 + R_2) + j(X_1 + X_2)$$

For complex impedances in parallel, the process of finding the net impedance is more complicated. We'll deal with it later in this chapter.

PROBLEM 3-5

Suppose that an inductor has reactance $X_L = 45$ ohms at a certain frequency, and a capacitor has reactance $X_C = -55$ ohms at that same frequency. If we connect these two components in series, what's the net impedance Z across the combination?

SOLUTION

The complex impedance of the inductor is $Z_L = 0 + j45$, and the complex impedance of the capacitor is $Z_C = 0 - j55$. When they're connected in series, we can calculate the net impedance Z as

$$Z = Z_L + Z_C$$
$$= (0 + j45) + (0 - j55)$$
$$= 0 + 0 + j(45 - 55)$$
$$= 0 + j(-10)$$
$$= 0 - j10$$

This complex number represents a pure capacitive reactance of –10 ohms.

PROBLEM 3-6

Imagine that the inductor in the above circuit changes so its reactance becomes $X_L = 55$ ohms, while the reactance of the capacitor remains the same at –55 ohms. Also suppose that the inductor has an internal DC resistance of 11 ohms. What's the net impedance Z across these two components connected in series?

SOLUTION

In this case, the impedance of the inductor equals $Z_L = 11 + j55$, and the impedance of the capacitor equals $Z_C = 0 - j55$. Therefore, the net series impedance Z is

$$Z = Z_L + Z_C$$
$$= (11 + j55) + (0 - j55)$$
$$= 11 + 0 + j(55 - 55)$$
$$= 11 + j0$$

We have a pure resistance of 11 ohms. The capacitive and inductive reactances cancel each other out, leaving us with no net reactance. Engineers refer to the condition of equal and opposite reactances as *resonance*. We call a series circuit that contains both inductance and capacitance, and in which $X_L = -X_C$, a *series-resonant circuit*.

Series-Resonant Frequency

When a series circuit contains inductance and capacitance, a condition of resonance always exists at a certain frequency. Let's represent the inductance (in henrys) as L, and let's call the capacitance (in farads) C. In that case, we can calculate the *series-resonant frequency* (in hertz), which we call f_o, using the formula

$$f_o = 1/[2\pi(LC)^{1/2}]$$
$$= [2\pi(LC)^{1/2}]^{-1}$$

This formula also holds when we express the value of f_o in megahertz, the value of L in microhenrys, and the value of C in microfarads.

 PROBLEM 3-7

Suppose that we connect a 100-µH inductor in series with an 0.0100-µF capacitor. What's the resonant frequency of this circuit, rounded to the nearest hundredth of a megahertz?

 **SOLUTION**

In this situation, $L = 100$ (in microhenrys) and $C = 0.0100$ (in microfarads), so we can use the above formula to calculate f_o directly in megahertz as

$$f_o = 1/[2\pi(100 \times 0.0100)^{1/2}]$$
$$= 1/(2\pi)$$
$$= 1/(2 \times 3.14159)$$
$$= 1/6.28318$$
$$= 0.16 \text{ MHz}$$

TIP *If we connect a pure resistance in series with the inductor and capacitor in a series LC circuit, the resulting circuit has the same resonant frequency as it would without the resistance. The resonant effect gets less pronounced as the resistance increases, but it never disappears unless the resistance rises to "infinity" (an open circuit).*

Admittance

Admittance quantifies the ease with which a medium carries AC. It constitutes the AC counterpart of DC conductance. We express admittance as a complex-number quantity, just as we do with impedance. Admittance has a real-number part that represents conductance, and an imaginary-number part that represents susceptance.

AC Conductance

In equations for AC circuits, we symbolize conductance as an italicized, uppercase letter G. The relationship between conductance and resistance is

$$G = 1/R$$

The unit of conductance is the *siemens* (called the *mho* in some archaic documents), abbreviated as a nonitalicized, uppercase letter S. As the conductance G increases, the resistance R goes down and the current I goes up. Conversely, as G decreases, R goes up and I goes down.

Susceptance

We symbolize susceptance in equations by writing an uppercase, italicized letter B. Mathematically, susceptance is the reciprocal of AC reactance. Susceptance, like reactance, can be either capacitive or inductive. In formulas and equations, we symbolize capacitive susceptance as B_C and inductive susceptance as B_L. The unit of susceptance is the siemens.

The complete theoretical expression of susceptance requires the j operator, just as does the expression of reactance. The j operator behaves strangely when it constitutes the denominator in a fraction or a quotient because the reciprocal of j equals its negative! That is,

$$1/j = -j$$

Therefore, the reciprocal of an imaginary number jX, where X represents the real-number multiple of j, is

$$1/(jX) = -j(1/X)$$

TIP *When we want to find the complex-number susceptance equivalent for a given complex-number reactance value, we must change the sign of j, and then take the reciprocal of the real-number component.*

If we let B_C represent the real-number component of the *capacitive susceptance* in siemens, f represent the frequency in hertz, and C represent the capacitance in farads, then

$$B_C = 2\pi f C$$

If we let B_L represent the real-number component of the *inductive suscep-tance* in siemens, f represent the frequency in hertz, and L represent the inductance in henrys, then

$$B_L = -1/(2\pi fL)$$
$$= -(2\pi fL)^{-1}$$

Putting Things Together

Admittance is the complex-number composite of conductance and suscep-tance. In that form, we can describe admittance Y as

$$Y = G + jB$$

If the j operator happens to be negative, we can write

$$Y = G - jB$$

In a parallel circuit, resistance and reactance don't combine neatly to yield impedance, but conductance and susceptance add together to give us the admittance. If we have two admittances $Y_1 = G_1 + jB_1$ and $Y_2 = G_2 + jB_2$ con-nected in parallel, the net admittance Y equals their sum

$$Y = (G_1 + G_2) + j(B_1 + B_2)$$

The *GB* Half-Plane

Engineers portray admittance values on a coordinate grid that looks superfi-cially like the complex impedance (RX) half-plane. Conductance values go along the horizontal, or G, axis, and susceptance values go along the B axis. Figure 3-18 shows this grid with several complex admittance points plotted. The center, or *origin*, of the *GB half-plane* represents the point at which there exists no conduction for DC or AC.

In a mathematical sense, the GB half-plane is "inside-out" compared with the RX half-plane, as follows:

- In the RX half-plane, the origin point represents a perfect short circuit, and points "infinitely far away" from the origin in all directions represent a perfect open circuit.

- In the GB half-plane, the origin point represents a perfect open circuit, and points "infinitely far away" from the origin in all directions represent a perfect short circuit.

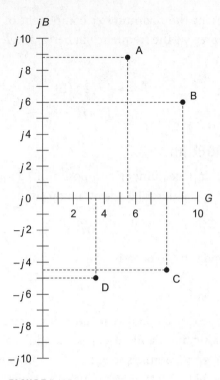

FIGURE 3-18 · Some points in the complex admittance (*GB*) half-plane. At A, 5.5 + *j*8.8. At B, 9 + *j*6. At C, 8 − *j*4.5. At D, 3.4 − *j*5.

In Fig. 3-19, we render the points from Fig. 3-18 as vectors. Longer vectors indicate relatively more current, and shorter vectors indicate relatively less current. Vectors pointing upwards and to the right represent conductances and capacitances in parallel. Vectors pointing downwards and to the right represent conductances and inductances in parallel.

Converting Admittance to Impedance

Imagine that you've found the net complex admittance of a parallel circuit as $G + jB$. If you want to know the net complex impedance of the circuit, you can find the resistance and reactance components individually using the formulas

$$R = G/(G^2 + B^2)$$

and

$$X = -B/(G^2 + B^2)$$

Once you've completed these calculations, you can express the complex impedance as the simple sum $R + jX$. Mathematically, impedance is the reciprocal of

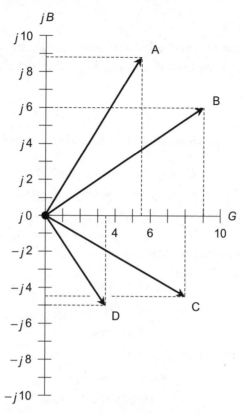

FIGURE 3-19 · Some vectors in the *GB* half-plane.
At A, $5.5 + j8.8$. At B, $9 + j6$. At C, $8 - j4.5$. At D, $3.4 - j5$.

admittance, but when you work with complex numbers, you get a more complicated situation than you do with "plain old real numbers."

When you're analyzing a parallel circuit containing resistance, inductance, and capacitance and you want to determine the complex impedance of the combination, go through the following steps in order:

- Find the conductance $G = 1/R$ for the resistor. It will be positive or zero.
- Find the susceptance B_L of the inductor using the appropriate formula. It will be negative or zero.
- Find the susceptance B_C of the capacitor using the appropriate formula. It will be positive or zero.
- Find the net susceptance $B = B_L + B_C$. It might be positive, negative, or zero.
- Compute R and X in terms of G and B using the appropriate formulas.
- Assemble the complex impedance $R + jX$.

PROBLEM 3-9

Suppose that the complex admittance of a certain parallel circuit is 0.010 $-j\,0.0050$. What's the complex impedance of this circuit if the frequency doesn't change?

SOLUTION

In this case, $G = 0.010$ S and $B = -0.0050$ S. First, we find $G^2 + B^2$, as follows:

$$G^2 + B^2 = 0.010^2 + (-0.0050)^2$$
$$= 0.000100 + 0.000025$$
$$= 0.000125$$

Next, we calculate the resistance R as

$$R = G/0.000125$$
$$= 0.010/0.000125$$
$$= 80 \text{ ohms}$$

and the reactance X as

$$X = -B/0.000125$$
$$= 0.0050/0.000125$$
$$= 40 \text{ ohms}$$

We conclude by assembling the complex impedance as the simple sum

$$R + jX = 80 + j\,40$$

PROBLEM 3-10

Suppose that we parallel-connect three components: a resistor whose conductance equals exactly 0.01 S, an inductor whose susceptance equals exactly -0.02 S, and a capacitor whose susceptance equals exactly 0.03 S. What's the complex admittance of the resulting circuit? What's its complex impedance?

SOLUTION

In order to solve this problem, we must go carefully through several steps and tolerate some tedious arithmetic, as follows:

- Note that $G = 0.01$ S.
- Note that $B_L = -0.02$ S.
- Note that $B_C = 0.03$ S.
- Determine $B = B_L + B_C = -0.02 + 0.03 = 0.01$ S.
- Assemble the complex admittance $G + jB = 0.01 + j0.01$.
- Determine $G^2 + B^2 = 0.01^2 + 0.01^2 = 0.0002$.
- Calculate $R = G/(G^2 + B^2) = 0.01/0.0002 = 50$ ohms.
- Calculate $X = -B/(G^2 + B^2) = -0.01/0.0002 = -50$ ohms.
- Assemble the complex impedance $R + jX = 50 - j50$.

TIP   *Sometimes you'll see circuits in which several resistors, capacitors, and/or inductors appear in series and parallel combinations. You can always reduce such a network to an equivalent series or parallel* resistance-inductance-capacitance (RLC) *circuit that contains, in effect, only one resistance, one capacitance, and one inductance. Resistances in series simply add up. Inductances in series also add up. Capacitances in series combine according to the formula*

$$C = 1/[1/C_1 + 1/C_2 + 1/C_3 + \ldots + 1/C_n]$$

where $C_1, C_2, \ldots,$ *and* C_n *represent the individual capacitances, and C represents the total (or net) capacitance. In parallel, resistances and inductances combine as capacitances do in series. Capacitances in parallel simply add up.*

PROBLEM 3-11

Suppose that we connect a resistor having a conductance of exactly 0.02 S, an inductor having a susceptance of exactly −0.04 S, and a capacitor having a susceptance of exactly 0.04 S in parallel. What's the complex admittance? What's the complex impedance?

SOLUTION

This solution proceeds like the solution to Prob. 3-10, with slightly different numbers (and a dramatically different outcome). Let's go:

- Note that $G = 0.02$ S.
- Note that $B_L = -0.04$ S.
- Note that $B_C = 0.04$ S.

- Determine $B = B_L + B_C = -0.04 + 0.04 = 0$ S.
- Assemble the complex admittance $G + jB = 0.02 + j0$.
- Determine $G^2 + B^2 = 0.020^2 + 0.00^2 = 0.00040$.
- Calculate $R = G/(G^2 + B^2) = 0.02/0.0004 = 50$ ohms.
- Calculate $X = -B/(G^2 + B^2) = 0/0.00040 = 0$ ohms.
- Assemble the complex impedance $R + jX = 50 + j0$.

In this case, we have a pure conductance of precisely 0.02 S, which corresponds to a pure resistance of precisely 50 ohms. The capacitive and inductive susceptances cancel each other out, leaving us with no net susceptance (and therefore no net reactance). We call a parallel circuit that contains both inductance and capacitance, and in which $B_L = -B_C$, a *parallel-resonant circuit*.

Parallel-Resonant Frequency

Parallel resonance always occurs at a defined frequency, which we can find using the same formula as the one that applies to series resonance, given earlier in this chapter. If we represent the inductance (in henrys) as L and the capacitance (in farads) as C, then we can find the *parallel-resonant frequency* (in hertz), f_o, using the formula

$$f_o = 1/[2\pi(LC)^{1/2}]$$
$$= [2\pi(LC)^{1/2}]^{-1}$$

As in the series case, this formula also holds when we express the value of f_o in megahertz, the value of L in microhenrys, and the value of C in microfarads.

TIP   *If we connect a pure resistor in series with either the inductor or the capacitor in a* parallel LC circuit *(that is, an inductor and a capacitor in parallel), the resulting circuit has the same resonant frequency as it would without the resistor. The resonant effect gets less pronounced as the resistance increases, but it never goes away entirely unless the resistance goes up to "infinity" (an open circuit). If we connect a pure resistor in parallel with the inductor and capacitor, the resonant effect gets less pronounced as the resistance decreases, but it never vanishes altogether unless the resistance goes all the way down to zero (a short circuit).*

QUIZ

Refer to the text in this chapter if necessary. A good score is eight correct. The answers are listed in the back of the book.

1. What's the sum of $5 + j7$ and $-3 - j2$?
   - A. $8 + j9$
   - B. $8 + j5$
   - C. $2 + j5$
   - D. $2 + j9$

2. What's the reciprocal of $j2$?
   - A. $j0.5$
   - B. $-j0.5$
   - C. $-j2$
   - D. We need more information to calculate it.

3. What's the net complex impedance of a 100-ohm resistor connected in series with a 100-μH inductor at a frequency of 318.31 kHz?
   - A. $100 + j200$
   - B. $100 - j200$
   - C. $100 + j50$
   - D. $100 - j50$

4. What's the net complex impedance of a 100-ohm resistor connected in series with a 1000-pF capacitor at a frequency of 3.1831 MHz?
   - A. $100 + j200$
   - B. $100 - j200$
   - C. $100 + j50$
   - D. $100 - j50$

5. What's the net complex impedance of a 100-ohm resistor, a 100-μH inductor, and a 1000-pF capacitor, all connected in series?
   - A. $100 + j0$
   - B. $100 + j150$
   - C. $100 + j50$
   - D. We need more information to say.

6. What's the resonant frequency of a circuit containing a 100-ohm resistor, a 100-μH inductor, and a 1000-pF capacitor, all connected in series?
   - A. 2.52 MHz
   - B. 1.99 MHz
   - C. 503 kHz
   - D. 252 kHz

7. What happens to the resonant frequency of the circuit described in Question 6 if we short out the resistor, leaving only the inductor and capacitor in series?

   A. It decreases.
   B. It increases.
   C. It stays the same.
   D. We need more information to say.

8. What's the net complex admittance of a 100-ohm resistor connected in parallel with a 100-µH inductor at a frequency of 318.31 kHz?

   A. $0.0100 + j0.00500$
   B. $0.0100 - j0.00500$
   C. $0.0100 + j0.0100$
   D. $0.0100 - j0.0100$

9. What's the net complex admittance of a 100-ohm resistor connected in parallel with a 1000-pF capacitor at a frequency of 3.1831 MHz?

   A. $0.0100 + j0.0200$
   B. $0.0100 - j0.0200$
   C. $0.0100 + j0.0500$
   D. $0.0100 - j0.0500$

10. What's the resonant frequency of a circuit containing a 100-ohm resistor, a 100-µH inductor, and a 1000-pF capacitor, all connected in parallel?

   A. 2.52 MHz
   B. 1.99 MHz
   C. 503 kHz
   D. 252 kHz

Power Supplies

Most electronic circuits require DC to work properly. In the United States, the electricity comes from utility providers as AC at a nominal frequency of 60 Hz and nominal voltages of either 117 V or 234 V RMS. A *power supply* converts the utility AC to DC having a predetermined voltage for a specific electronic device or system. Figure 4-1 is a block diagram of a typical power supply.

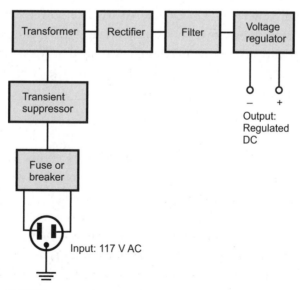

FIGURE 4-1 · Block diagram of a basic DC power supply.

CHAPTER OBJECTIVES

In this chapter you will

- See how transformers increase and decrease AC voltages.
- Analyze and compare half-wave and full-wave rectifier circuits.
- Learn how to filter and regulate rectifier output to obtain pure DC.
- See how to protect appliances from transients and other anomalies.
- Compare different types of electrochemical cells and batteries.
- Learn how power inverters and solar power supplies work.

Transformers

Power-supply transformers exist in two forms: the *step-down transformer* that decreases the AC voltage, and the *step-up transformer* that increases the AC voltage. The AC output frequency always equals the AC input frequency.

Step-Down Transformer

Most electronic devices, such as radio receivers, personal computers, and cell phones, need only a few volts of DC electricity. The power supplies for such equipment use step-down power transformers (Fig. 4-2A). The physical size of the transformer depends on the maximum current that we want it to deliver. The *input-to-output voltage ratio* depends directly on the *primary-to-secondary turns* ratio. Mathematically, we can say that

$$V_{in}/V_{out} = N_{pri}/N_{sec}$$

where V_{in} represents the RMS input voltage, V_{out} represents the RMS output voltage, N_{pri} represents the number of turns in the primary winding, and N_{sec} represents the number of turns in the secondary winding.

TIP   *The transformer in a low-current system, such as a radio receiver, can be physically small. But high-current devices, such as large audio amplifiers or amateur-radio transmitters, need bulky, heavy transformers. In a high-current system, the transformer secondary winding must comprise heavy-gauge (thick) wire, and the core, or material on which the wire is wound, must have considerable volume and mass.*

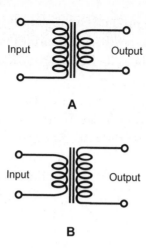

FIGURE 4-2 · At A, the schematic symbol for a step-down power transformer. At B, the schematic symbol for a step-up power transformer.

Step-Up Transformer

Some circuits require high voltages to work properly. For example, the *cathode-ray tube* (CRT) in an old-fashioned television receiver needs several hundred volts DC. Some amateur-radio *power amplifiers* use vacuum tubes working at several kilovolts DC. The transformers in these appliances are step-up types (Fig. 4-2B).

The voltage-transformation formula for a step-up transformer is the same as the formula given above for step-down transformers. Also, as in the step-down situation, the input-to-output voltage ratio equals the primary-to-secondary turns ratio.

TIP *If we need a step-up transformer to supply only a small current, it need not have great bulk or mass. However, some step-up transformers must operate "power-hungry" vacuum-tube equipment, such as broadcast transmitting amplifiers. For this reason, step-up transformers are generally larger and heavier than the step-down units in low-voltage systems, such as computers.*

Transformer Ratings

Engineers and manufacturers rate transformers according to their *output voltage* and their *maximum deliverable current*. For a given unit, the *volt-ampere*

(VA) *capacity*, which is the product of the output voltage and the maximum deliverable current, is often specified. A transformer that produces 12 V RMS of AC output and is capable of delivering 10 A RMS would have a rating of 12 × 10, or 120 VA.

The nature of power-supply filtering, discussed later in this chapter, makes it necessary for the power-transformer VA rating to be greater than the actual power, in watts, consumed by the load. A rugged power transformer, capable of providing the necessary current and/or voltage on a continuous basis, is crucial. The transformer usually constitutes the most expensive component in a power supply.

PROBLEM 4-1

Suppose you need a transformer that will provide 12 V RMS AC output when supplied with 120 V RMS AC input. Do you need a step-down type or a step-up type?

SOLUTION

Because the output voltage is lower than the input voltage, you need a step-down transformer.

PROBLEM 4-2

What primary-to-secondary turns ratio should the transformer have in the situation of Prob. 4-1?

SOLUTION

The primary-to-secondary turns ratio (N_{pri}/N_{sec}) must equal the input-to-output voltage ratio (V_{in}/V_{out}). You've been told that $V_{in} = 120$ V RMS and $V_{out} = 12$ V RMS. Therefore

$$V_{in}/V_{out} = 120/12$$
$$= 10$$

meaning that $N_{pri}/N_{sec} = 10$. Because this number represents a ratio, you can also write it as 10:1 (read "10 to 1").

Rectifiers

A *rectifier* converts AC to *pulsating DC*, usually by means of one or more heavy-duty *semiconductor diodes* following a power transformer.

Half-Wave Circuit

The simplest type of rectifier circuit, known as the *half-wave rectifier* (Fig. 4-3A), uses one diode (or a series or parallel combination of diodes if high current or voltage is required) to "chop off" half of the AC cycle. The effective voltage in this type of circuit equals roughly 35% of the peak voltage; Fig. 4-4A shows an example. The peak voltage in the reverse direction, called the *peak inverse voltage* (PIV) or *peak reverse voltage* (PRV) across the diode, can be up to 2.8 times the applied RMS AC voltage. Most engineers like to use diodes whose PIV ratings equal at least 1.5 times the maximum expected PIV. Therefore, with a half-wave supply, the diodes should be rated for at least 2.8 × 1.5, or 4.2, times the RMS AC voltage that appears across the secondary winding of the power transformer.

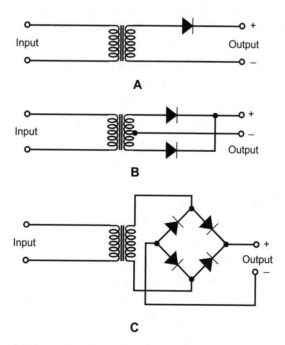

FIGURE 4-3 • At A, a half-wave rectifier circuit. At B, a full-wave center-tap rectifier circuit. At C, a full-wave bridge rectifier circuit.

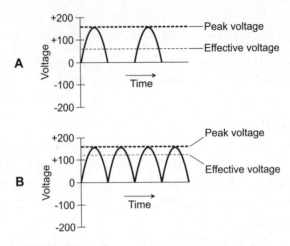

FIGURE 4-4 · At A, the output of a half-wave rectifier. At B, the output of a full-wave rectifier.

Half-wave rectification has shortcomings. First, the output can be difficult to filter, especially if the load demands high current. Second, the output voltage can diminish considerably when the supply delivers high current. Third, half-wave rectification puts a strain on the power transformer and the diodes because it "pumps" them. They can "loaf" during the half of the AC cycle when no current flows, but they must work hard during the half of the cycle when current does flow. Half-wave rectification usually works okay in a power supply that never has to deliver much current, or when the voltage can vary without affecting the behavior of the equipment connected to it. The main advantage of a half-wave circuit is its low co st compared with more sophisticated circuits.

Full-Wave Center-Tap Circuit

A better scheme for changing AC to DC takes advantage of both halves of the AC cycle. The so-called *full-wave center-tap rectifier* has a transformer with a tapped secondary, as shown in Fig. 4-3B. The center tap goes straight to *electrical ground*, so we get voltages and currents at the ends of the winding that occur in phase opposition with respect to each other. We individually half-wave rectify these two AC waves, alternately cutting off one half of the cycle and then the other, over and over.

In the circuit of Fig. 4-3B, the transformer center tap provides the negative DC voltage while the diodes provide the positive voltage. We can reverse the polarity of the DC output by reversing the orientations of the diodes. The effective voltage equals about 71% of the peak voltage, as shown in Fig. 4-4B. The PIV across the diodes can, as in the half-wave rectifier circuit, range up to 2.8 times the applied RMS AC voltage. Therefore, the diodes should have a PIV rating of at least 4.2 times the applied RMS AC voltage to ensure that they won't break down.

Full-Wave Bridge Circuit

In most applications, the *full-wave bridge rectifier*, sometimes called a *bridge* (if we know the context), offers the best method for converting AC to DC in a power supply. Figure 4-3C shows a schematic diagram of a full-wave bridge circuit. The output waveform resembles that of the full-wave center-tap circuit.

The effective output voltage in the bridge circuit equals about 71% of the peak voltage, as is the case with full-wave center-tap rectification. The PIV across the diodes equals approximately 1.4 times the applied RMS AC voltage. Therefore, each diode needs to have a PIV rating of at least 1.4×1.5, or 2.1, times the RMS AC voltage that appears at the transformer secondary.

The bridge circuit does not need a center-tapped transformer secondary. It uses the entire secondary winding on both halves of the wave cycle. The bridge circuit, therefore, makes more efficient use of the transformer than either the half-wave or the full-wave center-tap circuits do. The bridge circuit also treats the diodes more gently than either of the other rectifier arrangements do.

A bridge rectifier needs four diodes rather than two (in the case of a full-wave center-tap circuit) or one (in the case of a half-wave circuit). This increased complexity rarely amounts to much in terms of cost, but it makes a big difference when a power supply must deliver high current. Then, the extra diodes —two for each half of the cycle, rather than one—can dissipate more overall heat energy.

TIP *We can more easily filter the output of a full-wave rectifier than the output of a half-wave rectifier. The full-wave rectifier is also easier on the transformer and diodes than a half-wave circuit. If we connect a significant load to the output of the full-wave circuit, forcing it to deliver high current, the voltage drops less than it would with a half-wave supply under the same circumstances. In all these respects, the full-wave circuit outperforms the half-wave circuit.*

Voltage Multipliers

We can connect diodes and capacitors together to make a power supply deliver an output voltage at a whole-number multiple of the peak AC input voltage.

Theoretically, large whole-number multiples are possible, but we'll rarely see power supplies that make use of multiplication factors larger than 2.

In practice, *voltage-multiplier power supplies* work well only when the load draws low current. Otherwise, we'll get poor *voltage regulation*, meaning that the output voltage will go down quite a lot when the current demand rises significantly. In high-current, high-voltage applications, the best way to build a power supply involves the use of a step-up transformer, not a voltage-multiplication rectifier circuit.

Figure 4-5 shows a simple *voltage-doubler power supply*. This circuit works on the entire AC cycle, so engineers call it a *full-wave voltage doubler*. Its DC output voltage, when the current drain is low, equals roughly twice the peak AC in-put voltage, or about 2.8 times the applied RMS AC voltage. A full-wave voltage doubler subjects the diodes to a PIV of 2.8 times the applied RMS AC voltage. Therefore, the diodes should have a PIV rating of at least 4.2 times the RMS AC voltage that appears across the transformer secondary.

Proper operation of this type of circuit depends on the ability of the capacitors to hold a charge under maximum load. Therefore, the capacitors must have large values, as well as be capable of handling high voltages. The capacitors serve two purposes: to boost the voltage and to filter the output. The resistors, which have low ohmic values, protect the diode against *surge currents* that occur when we "switch the supply on," that is, when we initially apply AC power to the transformer.

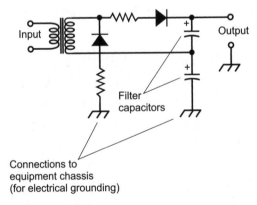

FIGURE 4-5 · A full-wave voltage-doubler power supply.

PROBLEM 4-3

Suppose that we connect a power transformer with a 1:4 primary-to-secondary turns ratio to the 120 V RMS AC utility mains. We use a half-wave rectifier circuit to obtain pulsating DC output. What's the minimum PIV rating the diodes should have in order to ensure that they won't break down?

SOLUTION

The utility mains supply 120 V RMS AC, and the transformer steps up this voltage by a factor equal to the primary-to-secondary turns ratio. Therefore, the output of the transformer equals 4 × 120 V RMS AC, or 480 V RMS AC. In a half wave circuit, the PIV rating of the diodes should be at least 4.2 times this value, so they must be rated for at least 4.2 × 480 PIV, or 2016 PIV.

PROBLEM 4-4

Suppose that we use a full-wave bridge rectifier circuit in the scenario of Problem 4-3. What's the minimum PIV rating the diodes should have in this case?

SOLUTION

The transformer secondary still delivers 480 V RMS AC. In a full-wave circuit, the PIV rating of the diodes should be at least 2.1 times the RMS AC voltage at the transformer secondary. That's half the PIV in the half-wave situation: 2.1 × 480 PIV, or 1008 PIV.

Filtering and Regulation

Most electronic equipment requires something better than the pulsating DC that comes straight out of a rectifier circuit. We can minimize or eliminate the pulsation, called *ripple*, in the rectifier output by means of a *power-supply filter*.

Capacitors Alone

The simplest power-supply filter consists of one or more large-value capacitors, connected in parallel with the rectifier output, as shown in Fig. 4-6. An *electrolytic capacitor* works well in this role. It's a *polarized* component, meaning that we must connect it in a certain direction, just as we would do with a battery or

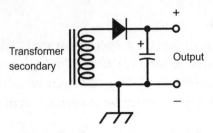

FIGURE 4-6 · A single capacitor can act as a power-supply filter.

a diode. If we connect an electrolytic capacitor the wrong way around, it won't work as a filter, and it might actually explode!

Typical electrolytic-capacitor values range from a few microfarads up to 100 µF or more. The more current that a power supply must deliver, the more capacitance we need for effective filtering because larger capacitances hold a charge for a longer time with a given current load than smaller capacitances do.

Filter capacitors work by "trying" to maintain the DC voltage at its peak level. This task is easier to carry out in the output of a full-wave rectifier (Fig. 4-7A) than in the output of a half-wave rectifier (Fig. 4-7B). With a full-wave rectifier receiving a 60-Hz AC electrical input, the ripple frequency equals 120 Hz; with a half-wave rectifier, the ripple frequency equals 60 Hz.

Still Struggling

The filter capacitor recharges twice as often with a full-wave rectifier, as compared with a half-wave rectifier. Therefore, we end up with less ripple at the filter output, for a given amount of capacitance, when we use a full-wave circuit.

Capacitors and Chokes

Another way to "smooth out" the DC from a rectifier involves placing a large-value inductor in series with the rectifier output, and a large-value capacitor in parallel with the rectifier output. The inductor, called a *filter choke*, normally has a value of several henrys.

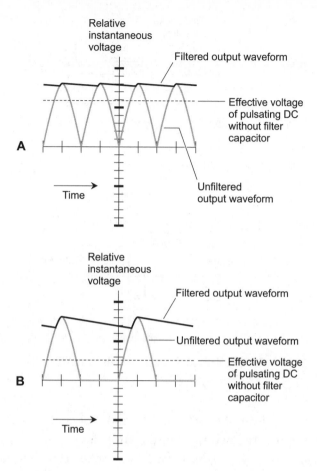

FIGURE 4-7 • Filtering of ripple in the output of a full-wave rectifier (A) and in the output of a half-wave rectifier (B).

If we place the capacitor on the rectifier side of the choke, we get a *capacitor-input filter* (Fig. 4-8A). If we place the choke on the rectifier side of the capacitor, we get a *choke-input filter* (Fig. 4-8B). We can use capacitor-input filtering when we don't expect or need a power supply to deliver high current. The output voltage is higher with a capacitor-input circuit than with a choke-input circuit. If we want our power supply to deliver large or variable amounts of current, we're better off using a choke-input filter because the output voltage will remain more stable.

If the output of a DC power supply must have an absolute minimum of ripple, we can connect two or three capacitor/choke pairs in *cascade*, as shown in Fig. 4-9. Each capacitor/choke pair constitutes a *section* of the filter. Multi-

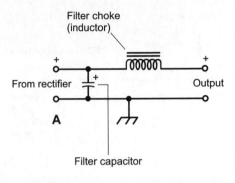

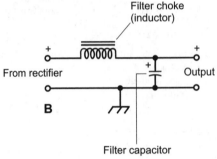

FIGURE 4-8 · At A, a capacitor-input filter. At B, a choke-input filter.

section filters can consist of either capacitor-input or choke-input sections, but we should never mix the two types in the same filter.

In the example of Fig. 4-9, both capacitor/choke pairs are called *L sections* because of their arrangement in the schematic diagram. If we omit the second capacitor, we get a *T section*. If we omit the first choke, we obtain a *pi section*.

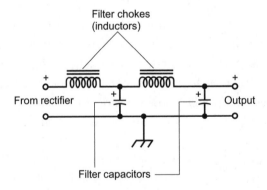

FIGURE 4-9 · Two choke-input filter sections in cascade.

These sections get their names from their appearances in schematic diagrams. If we use our imaginations, we can suppose that these component arrangements look like the uppercase English letter L, the uppercase English letter T, and the uppercase Greek letter Π respectively.

PROBLEM 4-5

Imagine that you live in a country where the standard AC line frequency is higher than 60 Hz. Does this frequency difference make the output of a rectifier circuit easier to filter, or more difficult, compared with the more familiar 60-Hz utility frequency?

SOLUTION

In theory, the higher AC frequency makes the output of a rectifier circuit easier to filter because the capacitors don't have to hold the charge for as long between cycle peaks.

Voltage Regulation

If we connect a specialized semiconductor component called a *Zener diode* in parallel with the output of a power supply along with a resistor in series, the Zener diode will limit the output voltage. The diode must have an adequate power rating to prevent it from burning out. The limiting voltage depends on the particular Zener diode used. We can find Zener diodes to fit any reasonable power-supply voltage.

Figure 4-10 is a diagram of a full-wave bridge DC power supply including a Zener diode for voltage regulation, along with a small-value resistor to limit the current through the diode. Note the direction in which the Zener diode is

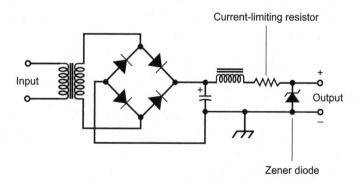

FIGURE 4-10 · A full-wave bridge power supply with a Zener-diode voltage regulator.

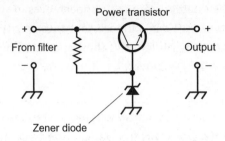

FIGURE 4-11 · A regulator circuit using a Zener diode and a power transistor.

connected: opposite from the way we would connect a rectifier diode! We must make sure that we connect the Zener diode in the right direction, or it will fail to regulate the voltage properly and will likely burn out.

A simple Zener-diode voltage regulator exhibits poor efficiency in a power supply that must deliver high current. When we need a power supply to deliver a lot of current, we can use a *power transistor* along with a Zener diode to obtain voltage regulation. Figure 4-11 is a schematic diagram of a regulator circuit that will work well with high-current power supplies.

Voltage regulators are available in *integrated-circuit* (IC) form. Such an IC, sometimes along with some external components, should be installed in the power-supply circuit at the output of the filter. In high-voltage power supplies, *electron-tube voltage regulators* are sometimes used.

Protecting Electronic Equipment

The output of a power supply should not exhibit sudden changes that can damage equipment or components, or interfere with their proper performance. Voltage must not appear on the external surfaces of a power supply, or on the external surfaces of any equipment connected to the supply.

Grounding

The best electrical ground for a power supply is the "third wire" ground provided in up-to-date AC utility circuits. The "third hole" (the bottom hole in an AC outlet, shaped like an uppercase English letter D turned on its side) should connect directly to a wire that ultimately terminates at a *ground rod* driven into the earth at the point where the electrical wiring enters the building.

In old buildings, *two-wire AC systems* are common. You can recognize this type of system by the presence of only two slots in the utility outlets. Some of these systems employ reasonable grounding by means of *polarization*, where one slot is longer than the other, and the longer slot goes to electrical ground. But that method is less effective than a *three-wire AC system*, in which the ground connection remains independent of both outlet slots.

TIP *Unfortunately, the presence of a three-wire or polarized outlet system doesn't always guarantee that an appliance connected to an outlet will be well grounded. If the appliance design is faulty, or if the "third hole" wasn't grounded by the people who installed the electrical system, a power supply can deliver a dangerous voltage to the external surfaces of appliances and electronic devices. This situation can present an electrocution hazard, and can also hinder the performance of electronic equipment.*

WARNING! *All metal chassis and exposed metal surfaces of AC power supplies should be connected to the grounded wire of a three-wire electrical cord. You should never defeat or cut off the "third prong" of the plug. You should find out whether or not the electrical system in the building was properly installed, so you don't work under the illusion that your system has a good ground when it actually does not. If you're in doubt about this issue, have a professional electrician perform a complete inspection of the system.*

Surge Currents

At the instant we switch a power supply on (usually by applying AC to the transformer primary by throwing a switch), a surge of current occurs, even with nothing connected to the supply output. The surge takes place because the filter capacitors need to acquire an initial charge, forcing them to draw a large current for a short time. The initial surge current can greatly exceed the normal operating current. An extreme surge can destroy the rectifier diodes in a power supply. This phenomenon sometimes presents a serious problem in poorly designed high-voltage power supplies and voltage-multiplier circuits. We can prevent diode failure as a result of current surges in at least four ways:

1. Use diodes with current and PIV ratings several times higher than the normal operating level.

2. Make certain that all the diodes are identical.

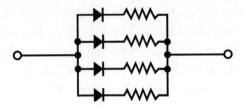

FIGURE 4-12 · Diodes in parallel, with current-equalizing resistors in series with each diode.

3. Connect several diodes in parallel wherever the circuit calls for a diode. *Current-equalizing resistors* of small ohmic value should be connected in series with each diode as shown in the hypothetical example of Fig. 4-12.

4. Use an automatic switching circuit in the transformer primary. A circuit of this type applies a reduced AC voltage to the transformer for the first second or two after the initial power-up, and then applies the full input voltage.

Transients

The AC on the utility line presents itself as a sine wave with a constant RMS voltage near 117 V or 234 V. But this wave is far from "pure"! If we look at the AC waveform on a high-quality laboratory oscilloscope, we'll occasionally see *voltage spikes*, known as *transients*, that attain peak values far higher than the peak waveform voltage. Transients result from sudden changes in the load in a utility circuit. A thundershower can produce transients throughout an entire town. Unless we take measures to suppress them, transients can destroy the diodes in a power supply. Transients can also interfere with the

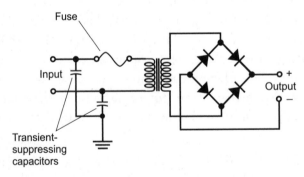

FIGURE 4-13 · A power supply with transient-suppression capacitors and a fuse in the transformer primary circuit.

operation of sensitive electronic equipment, such as computers or micro-computer-controlled appliances.

The simplest way to get rid of common transients involves placing a capacitor of about 0.01 μF, rated for 600 V or more, between each side of the transformer primary and electrical ground, as shown in Fig. 4-13. Commercially made *transient suppressors* are available. These devices, often mistakenly called "surge protectors," use specialized semiconductor-based components to prevent sudden voltage spikes from reaching levels where they can cause problems.

TIP *It's a good idea to use transient suppressors with all sensitive electronic devices, including computers, hi-fi stereo systems, and television sets. In the event of a thundershower, the best way to protect such equipment is to physically unplug it from the wall outlets until the storm has passed.*

Fuses

A *fuse* contains a piece of soft wire that melts, breaking a circuit if the current exceeds a certain level. We normally connect a fuse in series with the transformer primary, as shown in Fig. 4-13, along with transient-suppressing capacitors. A short circuit or overload will burn the fuse out. If a fuse blows out, we must replace it with another fuse having exactly the same current rating. Otherwise, we'll likely face either of two problems: frequent and unnecessary fuse burn-out (a mere nuisance) or inadequate equipment protection (a disaster waiting to happen)!

Fuses are available in two types: the *quick-break fuse* and the *slow-blow fuse*. A quick-break fuse has a straight length of wire or a metal strip. A slow-blow fuse usually has a spring inside along with the wire or strip. You should always replace blown-out fuses with new ones of the same type. Quick-break fuses in slow-blow situations might burn out needlessly (although in some cases they'll work okay). For example, a minor initial current surge, of no cause for concern in normal operation, can burn out a quick-break fuse. Slow-blow fuses in quick-break environments might not provide adequate protection to the equipment, letting excessive current flow for too long before burning out.

Circuit Breakers

A *circuit breaker* performs the same function as a fuse, except that we can easily reset a breaker by turning off the power supply, waiting a moment, and then pressing a button or flipping a switch. Some breakers reset automatically when the equipment has been shut off for a certain length of time.

If a fuse or breaker keeps blowing out or tripping, or if it blows or trips immediately after we've replaced or reset it, then trouble exists somewhere in the power supply or in the equipment connected to it. Possible problems include:

- One or more burned-out power-supply diodes
- A bad transformer
- Shorted filter capacitors
- A short circuit in the equipment connected to the supply
- A component connected with the wrong polarity

WARNING! *Never replace a fuse or breaker with a larger-capacity unit to overcome the inconvenience of repeated fuse/breaker blowing/tripping. Find the cause of the trouble, and repair the equipment as needed. The "penny in the fuse box" scheme can endanger equipment and personnel, and it increases the risk of fire in the event of a short circuit.*

PROBLEM 4-6

Will a transient suppressor work properly if it's designed for a three-wire electrical system but the ground wire has been defeated, cut off, or does not lead to a good electrical ground?

SOLUTION

No! In order to properly function, a transient suppressor needs a substantial ground. The device will divert excessive, sudden voltage "spikes" from sensitive equipment only when we provide it with a low-resistance current path to allow effective discharge to ground.

Electrochemical Power Sources

An *electrochemical cell*, often called simply a *cell*, is a source of DC power that can function independently of utility systems. When we connect two or more cells in series, we get a *battery*. Cells and batteries find applications in portable electronic equipment, in communications satellites, and as sources of emergency power.

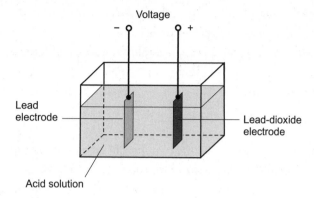

FIGURE 4-14 · A lead-acid electrochemical cell.

Electrochemical Energy

Figure 4-14 is a simplified drawing of a *lead-acid cell*. An electrode made of lead and another electrode made of lead dioxide, immersed in a sulfuric-acid solution, acquire a potential difference that can drive current through a load. The maximum available current depends on the volume and mass of the cell.

If we connect a lead-acid cell to a load for a long time, the current will gradually decrease, and the electrodes will become coated. The nature of the acid will change. All the potential energy in the acid will eventually end up as DC electrical energy, and then the cell will stop working. However, we can recharge it by connecting it to an external source of voltage equal to or a little greater than the normal cell voltage, positive-to-positive and negative-to-negative. The recharging time depends on the volume and mass of the cell, and can range from a few hours up to a day or more.

Primary and Secondary Cells

Some cells, once all their chemical energy has been changed to electricity and used up, must be discarded. We call these components *primary cells*. Other kinds of cells, such as the lead-acid unit described above, can get their chemical energy back again by recharging. We call this type of cell a *secondary cell*.

Primary cells contain a dry *electrolyte* paste along with metal electrodes. (The electrolyte contains the chemical energy that ultimately becomes DC electricity.) They go by names such as *dry cell*, *zinc-carbon cell*, or *alkaline cell*. We'll find them in supermarkets and department stores. Some secondary cells also show up in retail stores. *Lithium cells* are one common type. *Nickel-metal-hydride* (NiMH) cells are another. They cost more than ordinary dry cells, but these rechargeable cells can be reused hundreds of times, and can pay for themselves and the charger several times over.

An *automotive battery* comprises several secondary cells connected in series. The cells recharge from the vehicle's *alternator* or from an outside charging unit. A typical automotive battery contains lead-acid cells like the one shown in Fig. 4-14.

WARNING! *Never short-circuit the terminals of an automotive battery. Excessive current demand can cause the acid to boil out. In fact, it's unwise to directly short-circuit electrochemical cells or batteries of any type because under certain circumstances they can rupture, burn, or even explode.*

Standard Cell

Most cells produce between 1.0 and 1.8 V DC. Some types of cells generate predictable and precise voltages. These are known as *standard cells*. One example is the *Weston standard cell*, illustrated in Fig. 4-15. It produces 1.018 V DC

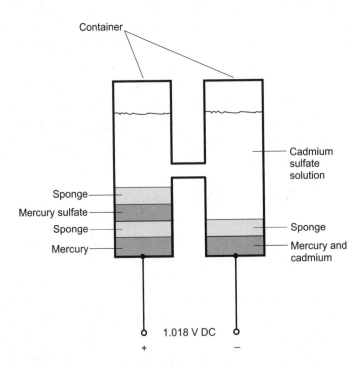

FIGURE 4-15 · A Weston standard electrochemical cell.

when operated at room temperature (approximately 20°C or 68°F). It has an electrolyte solution of cadmium sulfate. The positive electrode is mercury sulfate, and the negative electrode consists of mercury and cadmium. When a Weston standard cell is properly constructed and used at room temperature, its voltage is always the same. Engineers sometimes use it as a DC voltage reference.

Storage Capacity

In ordinary consumer applications, the most common units of electrical energy are the *watt-hour* (Wh) and the *kilowatt-hour* (kWh), where 1 kWh = 1000 Wh. Any cell or battery has a certain amount of electrical energy that can be specified in watt-hours or kilowatt-hours. Technically, engineers express the capacity of a cell or battery in terms of the mathematical *integral* of deliverable current with respect to time, in units of *ampere-hours* (Ah). The *storage capacity* in watt-hours equals the ampere-hour capacity multiplied by the battery voltage.

A battery with a rating of 20 Ah can provide 20 A for 1 h, or 1 A for 20 h, or 100 mA (100 milliamperes) for 200 h, or any of countless other demand characteristics. The extreme situations are the *shelf life* and the *maximum deliverable current*. Small cells have storage capacities on the order of a few milliampere-hours (mAh) up to 100 or 200 mAh. Medium-sized cells supply 500 mAh to 1000 mAh (1 Ah). Large automotive lead-acid batteries can provide upwards of 100 Ah.

Still Struggling

We define the shelf life as the length of time the battery will remain usable if we never connect it to a load; it can be months or years. We define the maximum deliverable current as the largest amount of current that a battery can drive through a load without a significant drop in its output voltage.

PROBLEM 4-7

Suppose that a battery can supply 10 Ah at 12 V DC. How many watt-hours (Wh) does this energy quantity constitute?

**SOLUTION**

Let's remember the basic formula for DC power in terms of voltage and current:

$$P = EI$$

where P represents the power in watts, E represents the voltage in volts, and I represents the current in amperes. Multiplying both sides by time t, in hours, gives us

$$Pt = EIt$$

In this equation, the left-hand side represents watt-hours, while the right-hand side represents the product of volts and ampere-hours. In our present situation, $E = 12$ and $It = 10$. Therefore

$$
\begin{aligned}
Pt &= EIt \\
&= 12 \times 10 \\
&= 120 \text{ Wh}
\end{aligned}
$$

Discharge Curves

When we place an *ideal cell* or *ideal battery* (a theoretically perfect unit that operates exactly according to a certain set of specifications) in service, we expect it to deliver a constant current for some time. After that, we expect the current to decrease. Some types of cells and batteries approach ideal behavior, exhibiting a *flat discharge curve* (Fig. 4-16A). Less optimal cells and batteries deliver current that decreases gradually from the beginning of use. That's a less desirable state of affairs, characterized by a *declining discharge curve* (Fig. 4-16B).

TIP   *When the current from a cell or battery has decreased to a level significantly less than its initial value, we say that the cell or battery is "weak" or "low." At that time, we should replace it. If we let a cell or battery "run down" until its output current drops to nearly zero, engineers and lay people might pronounce it "dead" (although, in the case of a rechargeable unit, a better term might be "completely discharged").*

Common Cells and Batteries

The cells sold in stores and used in convenience items, such as flashlights and transistor radios, are usually of the *zinc-carbon* or *alkaline* variety. These cells provide 1.5 V and are available in sizes designated as AAA (very small), AA (small), C

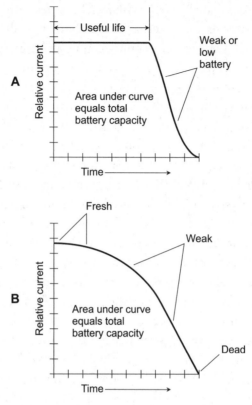

FIGURE 4-16 · At A, a flat discharge curve, representing the ideal for battery performance. At B, a declining discharge curve.

(medium), and D (large). Batteries made from zinc-carbon or alkaline cells usually produce 6 V (four internal cells in series) or 9 V (six internal cells in series).

Zinc-carbon cells have a fairly long shelf life. The zinc forms the outer case and acts as the negative electrode. A carbon rod serves as the positive electrode. The electrolyte comprises a paste of manganese dioxide and carbon. A zinc-carbon cell doesn't cost much money, and we can use it at moderate temperatures and in applications where the current drain is moderate to high. However, this type of cell doesn't work well at below-freezing temperatures.

Alkaline cells have granular zinc for the negative electrode, potassium hydroxide as the electrolyte, and a polarizer as the positive electrode. An alkaline cell can work at colder temperatures than a zinc-carbon cell. It also lasts longer when we use it to operate common electronic devices, and has a longer shelf life. Most people prefer alkaline cells to zinc-carbon cells for use in transistor radios, calculators, and portable cassette players.

Transistor batteries are small, 9-V, box-shaped batteries with clip-on connectors on top. They consist of six tiny internal zinc-carbon or alkaline cells connected in series. This type of battery finds application in low-current devices, such as wireless remote-control boxes, electronic calculators, and smoke detectors.

Lantern batteries are massive, and they can deliver a fair amount of current. One type has spring contacts on the top. The other type has thumbscrew terminals. Besides keeping an incandescent bulb lit for awhile, this type of battery, usually rated at 6 V and containing four internal zinc-carbon or alkaline cells, can provide enough energy to operate a small communications radio.

Silver-oxide cells have button-like shapes, and can fit inside a wristwatch. They come in various sizes and thicknesses, all with similar appearance. A silver-oxide cell supplies 1.5 V, and offers excellent energy storage capacity, considering its light weight. It also has a nearly flat discharge curve. Silver-oxide cells can be stacked to make batteries about the size of an AAA cylindrical cell.

Mercury cells, also called *mercuric oxide cells*, have advantages similar to silver-oxide cells. They are manufactured in the same general form. The main difference, often not of significance, is a somewhat lower voltage per cell: 1.35 V. In recent years, mercury cells (and batteries made from them) have lost their popularity because mercury presents an environmental hazard and is difficult to discard safely.

Lithium cells supply 1.5 V to 3.5 V, depending on the particular chemistry used in the manufacturing process. These cells, like their silver-oxide cousins, can be stacked to make batteries. Lithium cells and batteries have superior shelf life, and they can last for years in very-low-current applications. They offer excellent energy capacity per unit volume.

Lead-acid cells and batteries have a solution or paste of sulfuric acid, along with a lead electrode (negative) and a lead-dioxide electrode (positive). Paste-type lead-acid batteries can work well in consumer devices that require moderate current, such as laptop computers.

Nickel-Based Cells and Batteries

Nickel-cadmium (NICAD) cells and *nickel-metal-hydride (NiMH) cells* are made in several types. *Cylindrical cells* look like dry cells. *Button cells* are used in cameras, watches, memory backup applications, and other places where miniaturization is important. *Flooded cells* are used in heavy-duty applications and can have a storage capacity of as much as 1000 Ah. *Spacecraft cells* are made in packages that can withstand extraterrestrial temperatures and pressures. *NICAD batteries* and *NiMH batteries* are available in packs of cells that can be plugged into equipment to form part of the case for a device. An example is the battery pack for a handheld radio transceiver.

Nickel-based cells and batteries should never remain connected to a load after the current drops to zero. This state of affairs can cause the polarity of a cell (or of one or more cells in a battery) to reverse. Once polarity reversal occurs, a NICAD or NiMH cell or battery will no longer function properly, and we can legitimately call it "dead." For this reason, once a NICAD or NiMH cell or battery has lost its charge, we should remove it from service and recharge it straightaway.

NiMH cells and batteries are preferable to the older NICAD types because cadmium, found in NICAD units, exhibits human toxicity and can cause long-term environmental damage if not properly discarded. In this respect, cadmium resembles mercury. In some regions, government regulations mandate strict procedures for the disposal of mercury or NICAD cells and batteries.

 PROBLEM 4-8

Suppose that you want to use a battery to retain the memory contents of a microcomputer-controlled device for long periods during which you don't operate the hardware. This application requires very little current—almost none, in fact. You want the battery to have low mass, occupy a small volume, and require no maintenance. What type of battery will work well for this purpose?

 SOLUTION

A lithium battery will operate effectively in a situation of this sort. In fact, you'll often find lithium batteries in intermittent-use consumer devices, such as radio transceivers, portable computers, cell phones, and emergency lanterns, installed specifically for the purpose of microcomputer-memory backup.

Specialized Power Supplies

Some electronic systems employ specialized power sources, including *power inverters, uninterruptible power supplies, solar-electric energy systems,* and *fuel cells.* As people become increasingly aware of the importance of backup and emergency power systems, we should expect that these devices will become more popular.

Power Inverter

A power inverter, sometimes called a *chopper power supply*, is a circuit that delivers high-voltage AC, usually 117 V RMS, from a low-voltage DC source, usually 12 V to 13.5 V. Figure 4-17 is a simplified block diagram of a power inverter. The *chopper* consists of a low-frequency oscillator that opens and closes a high-current switching transistor. This process interrupts the battery current, producing pulsating DC. The transformer converts the pulsating DC to AC, and also steps up the voltage.

The output of a low-cost power inverter does not necessarily constitute a sine wave, but instead resembles a sawtooth or square wave. The frequency might be somewhat higher or lower than the standard 60 Hz. More sophisticated (and expensive) inverters produce fairly good sine-wave output and have a frequency close to 60 Hz. High-end power inverters are preferred (and in some cases required) for use with sensitive electronic equipment, such as computers and computer-controlled radios.

TIP *If we follow the transformer shown in Fig. 4-17 with a rectifier and filter, the device becomes a* **DC transformer,** *also called a* **DC-to-DC converter.** *A device of this type can provide hundreds of volts DC from a low-voltage battery.*

Uninterruptible Power Supplies

When we operate an electronic appliance from utility power, a system malfunction or failure can result from a *blackout, brownout, interruption*, or *dip*.

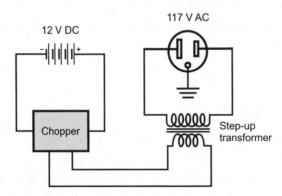

FIGURE 4-17 • A power inverter converts low-voltage DC into high-voltage AC.

We define a blackout as a complete loss of power for an extended period. A brownout represents a significantly reduced voltage for an extended period. An interruption constitutes a complete loss of power for a brief period. A dip is a significantly reduced voltage for a few moments.

To prevent power "glitches" from causing major trouble, such as computer data loss, we can use an *uninterruptible power supply* (UPS), such as the one diagrammed in Fig. 4-18. Under normal conditions, the equipment gets its power through the transformer and regulator. The regulator eliminates transients, surges, and dips in the utility power. A small current through the rectifier and filter maintains the lead-acid battery in a fully charged state.

If a utility power anomaly or failure occurs, an *interrupt signal* causes the switch to disconnect the equipment from the regulator and connect it to the power inverter, which converts the battery DC output to AC. For the duration of the utility "downtime," the battery discharges; its capacity should be sufficient to last long enough to allow for proper system shutdown. When utility power returns to normal, the switch disconnects the equipment from the battery and reconnects it to the regulator. Then the battery starts to charge up again.

TIP *If power to a computer fails and you have a UPS, save all your work immediately on the hard drive, and also on an external medium, such as a flash drive, if possible. Then switch the entire system, including the UPS, off until utility power returns, or until you've connected the system to a functioning emergency backup generator.*

Solar-Electric Power Supplies

We can obtain electricity directly from sunlight by means of *solar cells*, also known as *photovoltaic (PV) cells*. These devices, which constitute specialized semiconductor diodes, generate a few milliwatts of power for each square centimeter of surface area exposed to bright sunlight. Solar cells produce DC, while most household appliances require AC at 117 V RMS and 60 Hz. Most solar-electric energy systems intended for general home and business use must employ power inverters. Two main types of solar electric energy systems exist for providing electricity to homes and small businesses: the *stand-alone system* and *the interactive system*.

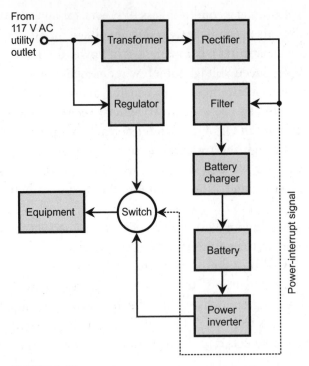

FIGURE 4-18 · An uninterruptible power supply (UPS) prevents system failure in case of utility power interruptions or irregularities.

A stand-alone photovoltaic system uses banks of rechargeable batteries, such as the lead-acid type, to store electric energy supplied by PV cells during daylight hours. The energy is released by the rechargeable batteries at night or in gloomy daytime weather. This system remains independent of the electric utility network (often called the *power grid*). An interactive system, also called a *grid-intertie system*, always remains connected to the power grid. It has no storage batteries. Excess energy, if any, is sold to utility companies during periods of bright daylight and minimum usage by means of a process called *net metering*. The conventional utility supplies energy at night, during gloomy daytime weather, or during times of heavy usage.

Fuel Cells

A *fuel cell* converts combustible gaseous or liquid fuel into usable electricity, but at a lower temperature than normal combustion. In practice, a fuel cell behaves like a battery that we can recharge by filling a fuel tank—or if we pipe the fuel in, by a continuous external supply. In the case of a fixed land location, the fuel can be either stored on-site or piped in. Some engineers have suggested conventional utility *methane* (also called *natural gas*) as the ideal fuel source for home power plants of this type because the delivery infrastructure already exists, and on-site storage is not necessary. However, in rural areas, or in any location not served by methane pipelines, other fuels such as hydrogen, alcohol, or gasoline can also work.

A typical fuel cell battery (called a *stack*) delivers several volts DC, comparable to the voltage produced by a solar array or automotive battery. Under normal conditions, the DC from the fuel cell goes to a power inverter that produces usable 117 V AC output. If desired, a backup battery bank can keep the electric current flowing when the fuel tank is refilled. A power control system switches the electrical appliances between the fuel cell and the battery bank as necessary. The DC output from the fuel cell or battery bank can also operate small appliances designed to run on low-voltage DC, such as two-way radios and notebook computers.

WARNING! *Power supplies can present a lethal electrical shock hazard. If you have any doubt about your ability to safely repair, modify, or maintain a power supply, leave the task to a professional technician.*

QUIZ

Refer to the text in this chapter if necessary. A good score is eight correct. The answers are listed in the back of the book.

1. If we apply standard 60-Hz, 120-V RMS AC to the primary winding of a transformer that has a primary-to-secondary turns ratio of exactly 2:1, what RMS AC voltage will we observe across the secondary winding?
   A. 480 V RMS
   B. 240 V RMS
   C. 60 V RMS
   D. 30 V RMS

2. If we connect four diodes in a full-wave bridge arrangement across the entire secondary of the transformer described in Question 1, what's the smallest allowable PIV rating for the diodes?
   A. 126 PIV
   B. 252 PIV
   C. 504 PIV
   D. 1008 PIV

3. Filter capacitors should have large values (in microfarads) to ensure that they can effectively
    A. handle high currents that might pass through them.
    B. eliminate the ripple in the rectifier output.
    C. prevent surge currents from damaging the transformer.
    D. keep avalanche breakdown from taking place in the diodes.

4. Which of the following component combinations can we connect at the output of a rectifier to serve as an effective filter?
    A. An inductor in parallel and a capacitor in series
    B. An inductor in series and a capacitor in parallel
    C. A single capacitor in series
    D. Any of the above

5. Which of the following measures reduces the risk of diode failure in a bridge rectifier as a result of the current surge that might occur when we apply AC to the transformer primary?
    A. Use four identical diodes.
    B. Use diodes with current ratings much higher than the anticipated operating current.
    C. Connect a small-value resistor in series with each diode.
    D. Any of the above

6. Imagine that a slow-blow fuse burns out in the power supply for your amateur radio transceiver as a result of rectifier-diode failure. You replace the faulty diodes with new ones having slightly higher PIV and current-handling ratings to reduce the chance of the same sort of failure happening again. However, you don't have any slow-blow fuses available; you have only quick-break types. You install a quick-break fuse with the same current and voltage ratings as the old slow-blow fuse. What should you expect?
    A. The new fuse *might* burn out even if no fault exists in the power supply or in the equipment connected to it.
    B. The new fuse *might* fail to burn out even if a problem occurs in the power supply or in the equipment connected to it.
    C. The new fuse will blow out immediately *every time* you switch on the power supply, whether a real fault exists or not.
    D. Any of the above

7. Which of the following cell or battery types *cannot* be recharged by connection to an external voltage source for a long time?
    A. A NiMH battery
    B. A lead-acid battery
    C. A zinc-carbon cell
    D. A secondary cell

8. Imagine that we need a battery that can provide 12 V DC at 1500 Ah to serve as the power supply for a communications unit at a location far away from any utility power source. We would do best to choose a battery comprising

   A. alkaline cells.
   B. NICAD cells.
   C. lithium cells.
   D. flooded cells.

9. So-called "transistor batteries" supply 9 V DC and contain

   A. six cells in parallel.
   B. six cells in series.
   C. nine photovoltaic cells in parallel.
   D. a solution of cadmium sulfate.

10. In order to provide electricity to a remote cabin in a location where no utility power service is available, we would most likely use

   A. a stand-alone PV system.
   B. an array of lithium cells.
   C. an uninterruptible power supply.
   D. a stack of Weston standard cells.

Test: Part I

Do not refer to the text when taking this test. You may draw diagrams or use a calculator if necessary. A good score is at least 38 correct. Answers are in the back of the book. It's best to have a friend check your score the first time, so you won't memorize the answers if you want to take the test again.

1. Suppose that we want to use a transformer to get 20 V RMS AC from a heavy-duty utility outlet that supplies 220 V RMS AC. The transformer should have a primary-to-secondary turns ratio of

   A. 121:1.
   B. 11:1.
   C. 1:1.
   D. 1:11.
   E. 1:121.

2. Ohm's law for DC circuits tells us that

   A. amperes equal volts divided by ohms.
   B. volts equal amperes times ohms.
   C. ohms equal volts divided by amperes.
   D. All of the above
   E. None of the above

3. Engineers occasionally use units called gilberts to express

   A. electric-field strength.
   B. electromotive power.
   C. magnetomotive force.
   D. charge-carrier acceleration.
   E. the ratio of power to energy.

4. In a pure resistance carrying an AC signal,

   A. the current and the voltage exist in phase coincidence.
   B. the current leads the voltage by 90°.
   C. the current lags the voltage by 90°.
   D. the current adds to the voltage.
   E. the current subtracts from the voltage.

5. If someone tells us that a certain AC signal has a frequency of 0.003320 MHz, we might prefer to express it as

   A. 3.320 kHz.
   B. 3320 GHz.
   C. 0.3320 THz.
   D. 33.20 Hz.
   E. 33,200 µHz.

6. What's the period of a wave whose frequency equals 440 Hz, rounded off to three significant figures? Remember that 1 ms = 0.001 s and 1 µs = 0.000001 s.

   A. 0.227 µs
   B. 2.27 µs
   C. 22.7 µs
   D. 0.227 ms
   E. 2.27 ms

7. **Which of the following three statements A, B, or C, if any, is true?**

    A. For a pure positive real number or 0, the absolute value equals the number itself.
    B. For a negative real number, the absolute value equals the number times −1.
    C. For a pure imaginary number $0 + jX$, the absolute value equals X when X is positive and $−X$ when X is negative.
    D. All of the above statements A, B, and C are true.
    E. None of the above statements A, B, or C is true.

8. **Engineers use a quantity called the j operator to define and calculate complex impedance values. Mathematically, the j operator equals**

    A. the diagonal of a square measuring 1 unit on each side.
    B. the circumference of a circle measuring 1 unit in radius.
    C. the number of degrees of phase in 1/4 of an AC cycle.
    D. the number of radians of phase in a full AC cycle.
    E. the positive square root of −1.

9. **Figure Test I-1 shows a circuit containing a DC battery, a DC voltmeter, a DC milliammeter, and a fixed resistor. What does the voltmeter indicate?**

    A. 11.0 V
    B. 22.0 V
    C. 110 V
    D. 114 V
    E. 220 V

10. **In the situation shown by Fig. Test I-1, how much power does the resistor dissipate?**

    A. 114 W
    B. 110 W
    C. 22.0 W
    D. 9.68 W
    E. 2.27 W

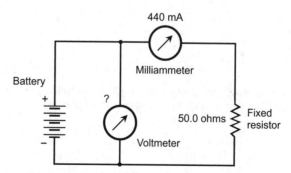

FIGURE TEST I-1 · Illustration for Part I Test Questions 9 and 10.

11. **In terms of physical units, an ampere represents**

    A. a potential change of 1 volt per second.
    B. a power change of 1 watt per second.
    C. a charge-carrier flow of 1 coulomb per second.
    D. an energy change of 1 joule per second.
    E. a temperature change of 1 degree Celsius per second.

12. **Suppose that an inductor has a reactance of 35 ohms at a certain frequency, and a capacitor has a reactance −80 ohms at that same frequency. If we connect these two components in series with a 68-ohm resistor, what's the net impedance of the whole circuit at the given frequency?**

    A. $23 + j45$
    B. $68 + j115$
    C. $68 - j45$
    D. $0 + j23$
    E. $23 - j45$

13. **An AC signal that contains all of its energy at a single frequency would appear as a**

    A. ramp wave on a frequency-domain display.
    B. sine wave on a time-domain display.
    C. square wave on an amplitude-domain display.
    D. sawtooth wave on a phase-domain display.
    E. All of the above

14. **In a pure capacitance carrying an AC signal,**

    A. the current and the voltage exist in phase coincidence.
    B. the current leads the voltage by 90°.
    C. the current lags the voltage by 90°.
    D. the current adds to the voltage.
    E. the current subtracts from the voltage.

15. **A typical full-wave center-tap rectifier circuit contains**

    A. one diode.
    B. two diodes.
    C. three diodes.
    D. four diodes.
    E. six diodes.

16. **Suppose that a span of utility line has an effective resistance of 20.0 ohms and carries 4.00 A of current. What's the power loss in the line?**

    A. 0.800 W
    B. 5.00 W
    C. 80.0 W
    D. 320 W
    E. We need more information to calculate it.

17. Consider two pure AC sinusoidal signals, neither of which has a DC component, both of which have the same peak-to-peak voltage, and both of which have the same frequency. How can we adjust these two waves so that they cancel each other out, leaving us with no net signal?

    A. We can adjust their phases so that they differ by 180°, leaving all other factors the same.
    B. We can adjust their frequencies so that one wave has twice the frequency of the other, leaving all other factors the same.
    C. We can adjust their amplitudes so that one wave has twice the peak-to-peak voltage of the other, leaving all other factors the same.
    D. We can add a DC component to one wave, and an equal but opposite DC component to the other, leaving all other factors the same.
    E. We can't!

18. What's the absolute value of the complex number 21.00 + j28.00, rounded off to four significant figures?

    A. 49.00
    B. 35.00
    C. 588.0
    D. 24.50
    E. 24.25

19. Suppose that an irregular wave has a positive peak voltage of +5.5 V pk+ and a negative peak voltage of −3.5 V pk−. What's the peak-to-peak voltage?

    A. 9.0 V pk-pk
    B. 4.5 V pk-pk
    C. 2.0 V pk-pk
    D. 1.0 V pk-pk
    E. We need more information to calculate it.

20. A typical full-wave bridge rectifier circuit contains

    A. one diode.
    B. two diodes.
    C. three diodes.
    D. four diodes.
    E. six diodes.

21. Scientists define permeability on a scale relative to the concentration of magnetic flux lines in

    A. iron.
    B. water.
    C. copper.
    D. silver.
    E. free space.

22. **What's the sum $(11 + j15) + (13 - j17)$?**
    A. $24 - j2$
    B. $24 + j32$
    C. $26 + j30$
    D. $-24 + j2$
    E. $-2 + j32$

23. **Figure Test I-2 shows two pure AC sine waves called P and Q, neither having a DC component. It appears that**
    A. waves P and Q differ in peak-to-peak amplitude.
    B. waves P and Q differ in frequency.
    C. wave P leads wave Q in phase.
    D. wave P lags wave Q in phase.
    E. more than one of the above are true.

24. **If each horizontal division in Fig. Test I-2 represents precisely 1 ms (0.001 s), what's the frequency of wave P, rounded off to four significant figures?**
    A. 1000 Hz
    B. 333.3 Hz
    C. 166.7 Hz
    D. 83.33 Hz
    E. We need more information to answer this question.

25. **Figure Test I-3 shows four possible vector representations for the situation illustrated in Fig. Test I-2. The signals show up as vectors called P (for wave P) and Q (for wave Q).**

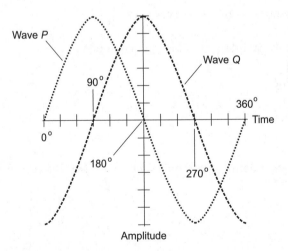

FIGURE TEST I-2 • Illustration for Part I Test Questions 23 and 24.

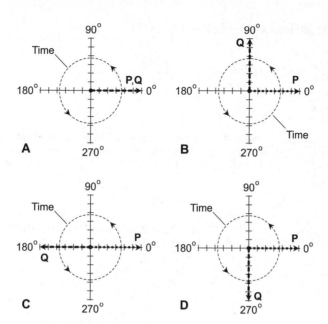

FIGURE TEST I-3 · Illustration for Part I Test Question 25.

As time passes, both vectors rotate counterclockwise at equal, constant angular speeds. Which vector graph portrays the situation correctly?

A. Graph A

B. Graph B

C. Graph C

D. Graph D

E. None of them

26. What's the frequency of a sine wave whose period equals 250 nanoseconds (ns), where 1 ns = 10^{-9} s?

A. 400 MHz

B. 40.0 MHz

C. 4.00 MHz

D. 400 kHz

E. 40.0 kHz

27. Suppose that an incandescent bulb draws 500 mA when connected to a lantern battery that supplies 6.30 V DC. What's the resistance of the bulb in this situation?

A. 3.15 ohms

B. 6.30 ohms

C. 12.6 ohms

D. 79.4 ohms

E. We need more information to calculate it.

28. **What's the difference $(11 + j15) - (13 - j17)$?**

   A. $24 - j2$
   B. $24 + j32$
   C. $26 + j30$
   D. $-24 + j2$
   E. $-2 + j32$

29. **Consider an inductor and a capacitor connected in parallel. If we represent the inductance (in microhenrys) as L and the capacitance (in microfarads) as C, then we can find the resonant frequency of the combination (in megahertz), f_o, using the formula**

$$f_o = 1/[2\pi(LC)^{1/2}]$$

   where π equals approximately 3.14159. Imagine a 100-μH inductor in parallel with a 500-pF capacitor. What's the resonant frequency of the combination?

   A. 1.40 MHz
   B. 712 kHz
   C. 3.18 MHz
   D. 314 kHz
   E. 628 kHz

30. **Figure Test I-4 illustrates the phase relationship between current and voltage in a series AC circuit containing**

   A. pure inductance.
   B. pure capacitance.
   C. pure resistance.
   D. inductance and resistance.
   E. capacitance and resistance.

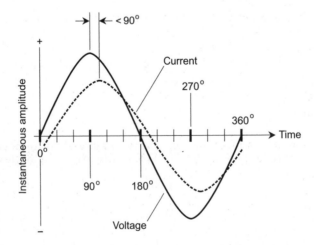

FIGURE TEST I-4 • Illustration for Part I Test Question 30.

31. **In practice, voltage-doubler power supplies work best when**
    A. the input voltage is extremely high or variable.
    B. the load resistance fluctuates over a wide range.
    C. the load does not draw much current.
    D. excellent voltage regulation is required.
    E. the load draws a large current.

32. **Suppose that we observe a DC potential difference of 24.0 V across a resistor that carries 250 mA of current. What's the value of the resistor?**
    A. 1.50 ohms
    B. 6.00 ohms
    C. 10.4 ohms
    D. 48.0 ohms
    E. 96.0 ohms

33. **Suppose that a battery can supply 70 Ah at 14 V DC. How much energy is that?**
    A. 0.98 kWh
    B. 42 Wh
    C. 31 Wh
    D. 5.0 Wh
    E. We need more information to calculate it.

34. **Nickel-metal-hydride (NiMH) cells are preferable to nickel-cadmium (NICAD) cells because**
    A. NiMH cells cost far less than NICAD cells.
    B. NiMH cells are less likely than NICAD cells to explode when recharged.
    C. the hydrogen in NiMH cells can be easily replenished.
    D. the cadmium in NICAD cells presents an environmental hazard, but NiMH cells contain no toxic elements.
    E. batteries made with NiMH cells can provide higher voltages than batteries made with NICAD cells.

35. **In a pure inductance carrying an AC signal,**
    A. the current and the voltage exist in phase coincidence.
    B. the current leads the voltage by 90°.
    C. the current lags the voltage by 90°.
    D. the current adds to the voltage.
    E. the current subtracts from the voltage.

36. **Which of the following devices constitutes a DC power source that we can recharge with methane (also called natural gas)?**
    A. Fuel cell
    B. Gas-discharge cell
    C. Tertiary cell
    D. Vapor cell
    E. None of the above; no such thing exists.

37. Consider again the situation described in Question 29, except that instead of connecting the 100-μH inductor and the 500-pF capacitor in parallel, we connect them in series. What's the resonant frequency of this new combination?

    A. 1.40 MHz
    B. 712 kHz
    C. 3.18 MHz
    D. 314 kHz
    E. 628 kHz

38. Suppose that we wind a length of wire 75 times around a plastic hoop in free space. Then we connect a battery to the coil and observe a current of 200 mA. What's the resulting magnetomotive force?

    A. 1875 At
    B. 375 At
    C. 75 At
    D. 15 At
    E. 3.0 At

39. If we insert a resistor into the circuit in the situation described by Question 38 so that the coil carries only 1/16 as much current as it did before, what will happen to the magnetomotive force?

    A. It will not change.
    B. It will decrease to 1/2 its former value.
    C. It will decrease to 1/4 of its former value.
    D. It will decrease to 1/8 of its former value.
    E. It will decrease to 1/16 of its former value.

40. Figure Test I-5 illustrates a power supply designed to convert conventional utility AC to constant, pure DC. What, if anything, is wrong with this circuit diagram?

    A. Nothing is wrong with this diagram.
    B. The Zener diode is connected backwards.
    C. One of the rectifier diodes is connected backwards.
    D. The filter capacitor is connected backwards.
    E. The filter choke must precede the capacitor, not follow it.

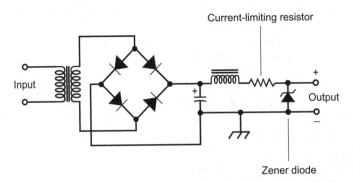

FIGURE TEST I-5 · Illustration for Part I Test Questions 40 and 41.

41. In the circuit of Fig. Test I-5, which of the following values might we reasonably expect the filter capacitor to have?

   A. 10 pF
   B. 100 pF
   C. 0.01 μF
   D. 0.1 μF
   E. 100 μF

42. A microvolt represents the equivalent of

   A. 10^{-9} kV.
   B. 10^{-6} kV.
   C. 10^{-3} kV.
   D. 10^{6} kV.
   E. 10^{9} kV.

43. If we increase the frequency of a sine wave by a factor of 36, what happens to its period?

   A. It decreases to 1/1296 of its former value.
   B. It decreases to 1/36 of its former value.
   C. It decreases to 1/6 of its former value.
   D. It increases by a factor of 6.
   E. It increases by a factor of 36.

44. Consider again the situation described in Question 37, where we have a 100-μH inductor in series with a 500-pF capacitor. Suppose that we add a 100-ohm resistor to the circuit between the inductor and the capacitor, in series with them both. What's the resonant frequency of the whole system now?

   A. 1.40 MHz
   B. 712 kHz
   C. 3.18 MHz
   D. 314 kHz
   E. No resonant frequency exists; the added resistance eliminates the resonant properties of the system.

45. To prevent a sudden power blackout from causing data loss in a desktop computer operating from a standard utility outlet, we can take advantage of

   A. an uninterruptible power supply.
   B. a full-wave bridge rectifier.
   C. a voltage-doubler power supply.
   D. a set of Weston standard cells.
   E. a grid-intertie photovoltaic system.

46. The angular frequency of an AC wave, in radians per second, equals

   A. $1/\pi$ times the frequency in hertz.
   B. π times the frequency in hertz.
   C. 2π times the frequency in hertz.
   D. $1/(2\pi)$ times the frequency in hertz.
   E. the frequency in hertz.

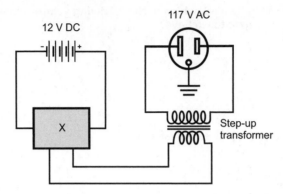

FIGURE TEST I-6 • Illustration for Part I Test Question 47.

47. Figure Test I-6 is a simplified diagram of a power inverter. What function, or set of functions, does the component marked X perform?
    A. It interrupts the battery current, producing pulsating DC.
    B. It eliminates variations in the battery voltage.
    C. It prevents excessive current drain from the battery.
    D. It ensures that the battery discharges gradually.
    E. All of the above

48. If we double the ampere-hour capacity of a storage battery and double its voltage as well, then its energy-storage capacity
    A. remains the same.
    B. increases by a factor of the square root of 2.
    C. increases by a factor of 2.
    D. increases by a factor of 4.
    E. increases by a factor of 8.

49. Suppose that we connect a 220-ohm resistor in series with a 330-ohm resistor. What's the net resistance of the combination?
    A. 550 ohms
    B. 275 ohms
    C. 269 ohms
    D. 132 ohms
    E. 110 ohms

50. Suppose that we connect a 220-ohm resistor in parallel with a 330-ohm resistor. What's the net resistance of the combination?
    A. 550 ohms
    B. 275 ohms
    C. 269 ohms
    D. 132 ohms
    E. 110 ohms

Part II

Wired Electronics

chapter **5**

Semiconductor Diodes

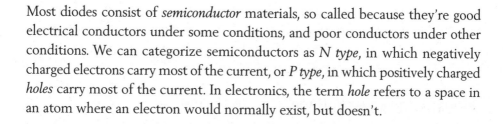

Most diodes consist of *semiconductor* materials, so called because they're good electrical conductors under some conditions, and poor conductors under other conditions. We can categorize semiconductors as *N type*, in which negatively charged electrons carry most of the current, or *P type*, in which positively charged *holes* carry most of the current. In electronics, the term *hole* refers to a space in an atom where an electron would normally exist, but doesn't.

CHAPTER OBJECTIVES

In this chapter you will

- Analyze semiconductor behavior, and learn how a P-N junction works.
- See how diodes can convert AC to DC.
- Learn how to use diodes for amplitude limiting.
- Compare diode-based signal mixers and frequency multipliers.
- Find out how engineers use diodes as oscillators and amplifiers.
- Discover diodes that convert radiant energy to DC or vice-versa.

The P-N Junction

When we place a *wafer* (small piece) of N type semiconductor material into direct physical contact with a wafer of P type semiconductor material, we get a *P-N junction* with unique and useful properties. A device with a single P-N junction has two electrodes, forming a *semiconductor diode*. Figure 5-1 shows the schematic symbol for this device. We represent the *cathode* (the electrode connected to the N type wafer) as a straight line, and the *anode* (the electrode connected to the P type wafer) as an arrow. With exceptions that we'll learn about shortly, electrons can flow easily from N to P in the direction opposite the arrow, but hardly at all from P to N in the direction that the arrow points.

Forward Bias

If we place a battery and resistor in series with a diode, current flows if we connect the negative terminal of the battery to the cathode and the positive terminal to the anode. In this situation, we have a condition called *forward bias* (Fig. 5-2A). In the N type material, electrons repel from the negative charge pole toward the P-N junction. In the P type material, holes repel from the positive charge pole toward the P-N junction. Therefore, electrons (in the N type material) and holes (in the P type) congregate near the junction, easily crossing it and producing current through the diode.

Forward Breakover Voltage

When we forward-bias a diode, we must subject the P-N junction to a certain minimum voltage to cause conduction. Engineers call this threshold the *forward breakover voltage*. Depending on the semiconductor elements used in the diode-manufacturing process, the forward breakover voltage can have a value

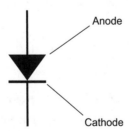

FIGURE 5-1 · Anode and cathode symbology for a semiconductor diode.

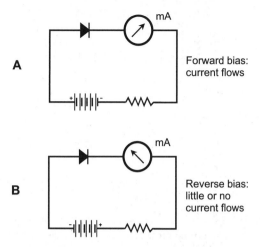

FIGURE 5-2 • Simplified description of diode behavior. At A, forward bias allows current to flow. At B, reverse bias practically prevents current from flowing.

anywhere between about 0.3 V and 1 V. In a *silicon diode*, the most common type of semiconductor diode used in power supplies, the forward breakover voltage is approximately 0.6 V. If the voltage across the junction doesn't equal or exceed the forward breakover voltage, the diode won't conduct.

Reverse Bias

If we connect the battery so that the positive terminal goes to the cathode and the negative terminal goes to the anode, we obtain *reverse bias* (Fig. 5-2B). In the N type material, electrons move toward the positive charge pole, away from the P-N junction. In the P type material, holes move toward the negative charge pole, also away from the junction. Therefore, the electrons and holes practically disappear in the vicinity of the junction, forming a *depletion region* that behaves as a thin *dielectric* layer, the equivalent of an electrical insulator. In this situation, little or no current can flow through the diode.

Avalanche Effect

If we reverse-bias a P-N junction and then increase the voltage without limit, we'll eventually reach a threshold voltage at which the diode suddenly begins to conduct. The P-N junction's insulating properties will succumb to the overwhelming EMF, charge carriers will "stampede" through the semiconductor material, and the depletion region will vanish. Engineers call this phenomenon

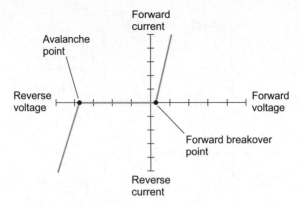

FIGURE 5-3 · Characteristic curve for a hypothetical diode, showing the forward breakover point and the avalanche point.

the *avalanche effect*, and the threshold voltage the *avalanche voltage*. The avalanche voltage varies among different kinds of diodes. Figure 5-3 is a graph of the current-versus-voltage *characteristic curve* for a typical semiconductor diode, showing the *forward breakover point* and the *avalanche point*.

Zero Bias

If we connect a diode in an electrical circuit so that no potential difference is imposed between the anode and the cathode, then we have a condition called *zero bias*. Obviously, a diode doesn't conduct any current in this situation!

Still Struggling

In the circuits of Fig. 5-2, the series-connected resistors, called *current-limiting resistors*, ensure that the current can't destroy the diode when the battery voltage exceeds either the forward breakover value or the avalanche value. In order to work properly, this resistance must be large enough to keep the flow of charge carriers below the maximum safe current specified by the diode manufacturer.

Junction Capacitance

Under conditions of reverse bias at a voltage below the avalanche point, a P-N junction can act as a capacitor. Manufacturers produce *varactor diodes* with this property specifically in mind. (The term is a contraction of the technical expression *variable reactor*, reflecting the fact that the component exhibits variable reactance.) Most varactors are made from silicon, or from a compound called *gallium arsenide* (GaAs).

In a varactor, we can vary the *junction capacitance* by changing the reverse-bias voltage because this voltage affects the width of the depletion region. As the reverse voltage increases, the depletion region grows wider, and the capacitance goes down—until, of course, the reverse bias reaches the avalanche threshold. Then the capacitor in effect "shorts out," and the diode no longer exhibits variable junction capacitance.

Rectifier Diodes

A *rectifier diode* passes current in only one direction under normal operating conditions. This one-way conduction property makes a diode useful for changing AC to DC. You learned about this *rectification* process in Chapter 4. When we use diodes as rectifiers, we never want to see the avalanche effect because it degrades the efficiency of a power supply. We can prevent unwanted avalanche effect in a power supply by choosing diodes with avalanche voltage ratings significantly higher than the *peak inverse voltage* (PIV) to which the transformer secondary subjects them.

TIP *In the design of an extremely high-voltage power supply, we might have to use series combinations of diodes to prevent an avalanche effect. When we place two or more diodes in series, the voltage across them divides up equally between the individual components.*

Zener Diodes

Most diodes have avalanche voltages far higher than the reverse bias ever gets. The value of the avalanche voltage depends on how a diode is manufactured. *Zener diodes* are "tailored" to exhibit well-defined, constant avalanche voltages. A Zener diode takes advantage of the avalanche effect, turning it from a nuisance into a benefit.

Suppose that a certain Zener diode has an avalanche voltage, also called the *Zener voltage*, of 50 V. If we apply a reverse bias voltage to the P-N junction, the diode acts as an open circuit below 50 V. When the voltage reaches 50 V, the diode conducts. If we try to increase the reverse bias voltage past 50 V, more current flows through the P-N junction. If we place a low-to-moderate-value resistor between the Zener diode and the power supply, the Zener diode effectively prevents the reverse voltage from exceeding 50 V. (If we don't use the current-limiting resistor, the power supply will drive a destructive current through the diode.)

Still Struggling

We can find more sophisticated ways to get voltage regulation besides the use of Zener diodes and current-limiting resistors, but Zener diodes often provide the simplest and least expensive alternative. Zener diodes are available with a wide variety of voltage and power-handling ratings. Power supplies for low-current electronic devices commonly employ Zener-diode regulators.

Thyrectors

A *thyrector* is a semiconductor device that resembles two diodes connected in series with their polarities reversed. We can protect sensitive electronic equipment against the effects of sudden, dramatic spikes in the instantaneous voltage by connecting a thyrector between each of the AC power lines and electrical ground (Fig. 5-4). Such "spikes," which last for only a tiny fraction of an AC cycle but can reach several hundred volts, are called *transients*.

A thyrector normally acts as an open circuit. It doesn't conduct until the instantaneous voltage reaches a certain value. If the voltage across the thyrector exceeds the critical level, even for a few millionths of a second, then the

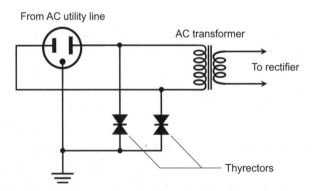

FIGURE 5-4 · Thyrectors can suppress transients in an AC line.

thyrector conducts, in effect "shorting out" transients by keeping the peak voltage from exceeding the critical level, called the *clipping voltage*.

Thyrectors can protect against most transients, but not against them all. A severe voltage spike, such as the *electromagnetic pulse* (EMP) from a nearby lightning strike, can destroy a thyrector and the equipment it's supposed to protect. The circuit of Fig. 5-4 constitutes a simple *transient suppressor*. In most department and hardware stores, you can find multioutlet AC power strips with built-in transient suppressors. They're often advertised as "surge protectors" or "surge suppressors," a technical misnomer.

PROBLEM 5-1

What's the difference between forward bias, reverse bias, and zero bias in a diode?

SOLUTION

In forward bias, the cathode (N type material) has a negative voltage relative to the anode (P type material). In reverse bias, the cathode has a positive voltage with respect to the anode. In zero bias, no externally applied voltage exists between the two electrodes.

PROBLEM 5-2

How does the avalanche voltage of a typical diode compare with the forward breakover voltage?

SOLUTION

We define avalanche voltage for reverse bias (cathode positive with respect to anode), while we define forward breakover voltage for forward bias (anode positive with respect to cathode). In most diodes, the avalanche voltage greatly exceeds the forward breakover voltage.

Signal Applications

Engineers use semiconductor diodes in a vast variety of audio-frequency (AF) and radio-frequency (RF) circuits to process electrical signals. The following sections outline some common applications of so-called *signal diodes*.

Envelope Detection

The first diode ever devised was a semiconductor device. Known as a *cat whisker*, it comprised a fine piece of wire in contact with a small piece of the mineral substance *galena*. The junction between the wire and the galena "crystal" acted as a *detector*, the equivalent of a rectifier for small RF currents. When experimenters connected the cat whisker in a circuit like the one shown in Fig. 5-5, they got a so-called "crystal-set" radio receiver capable of picking up *amplitude-modulated* (AM) signals.

You can use a special RF diode to build a "crystal-set" receiver. It works without any external source of power other than the signal itself. The diode recovers the audio from the AM radio signal, a process known as *envelope detection*. If you want the detector to work, the diode must function as a rectifier at RF, but not as a capacitor. Some modern RF diodes, called *point-contact diodes*, resemble microscopic versions of the cat whisker. The small contact area minimizes the capacitance.

Electronic Switching

The ability of diodes to conduct when forward-biased, and to insulate when reverse-biased, makes them useful for *electronic switching* in some applications. Diodes can switch signals much faster than any mechanical device can do.

One type of diode, made for use as an RF switch, has a specialized semiconductor layer sandwiched in between the P type and N type material. This layer, called an *intrinsic semiconductor*, reduces the junction capacitance so that the device can operate at higher frequencies than an ordinary diode does. The

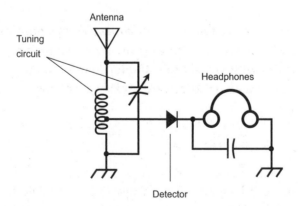

FIGURE 5-5 · A diode can serve as the envelope detector in a "crystal-set" radio receiver.

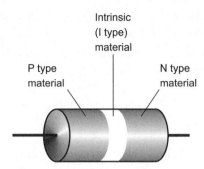

FIGURE 5-6 · A PIN diode has a thin layer of intrinsic (I type) material between the P and N type semiconductor layers.

intrinsic material is called *I type* material. A diode with an I type semiconductor layer between the cathode and the anode is called a *PIN diode* (Fig. 5-6).

If we apply DC bias to one or more PIN diodes, we can "channel" RF currents without using complicated, cumbersome relays and cables. The electronic switch appears closed when the diode has forward bias, and appears open when the diode has reverse bias.

Frequency Multiplication

When an AC signal passes through a semiconductor diode, the shape of the output wave differs considerably from the shape of the input wave. We call this effect *nonlinearity*. It results in an output signal rich in *harmonics* (signals whose frequencies equal whole-number multiples of the input signal frequency) and therefore renders diodes useful for *frequency multiplication*.

Figure 5-7 shows a simple frequency-multiplier circuit using a diode connected in series with the signal path. We tune the output inductance-capacitance

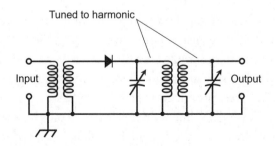

FIGURE 5-7 · We can use a diode to generate harmonics in a frequency-multiplier circuit.

(*LC*) circuits to resonate at the desired harmonic frequency rather than at the input or *fundamental frequency*. For example, if we want to multiply a 5-MHz input signal by a factor of 3, we should set both capacitors in the output circuit so that they, in conjunction with the transformer windings, cause resonance to occur at 15 MHz.

Step-Recovery Diodes

A *step-recovery diode*, also called a *charge-storage diode* or *snap diode*, is a specialized device that works well as a *harmonic generator*. When we drive an RF signal through a step-recovery diode, hundreds or thousands of harmonic signals appear at the output. Under forward-bias conditions, current flows in the same manner as it would in any forward-biased P-N junction. When we reverse the bias, the conduction continues for a short while, and then it abruptly ceases.

The step-recovery diode stores a large number of charge carriers while forward-biased. It takes a short time for these charge carriers to drain off after we reverse the bias. Current flows in the reverse direction until the charge carriers nearly disappear from the vicinity of the P-N junction, and then the current falls to zero almost instantly. The transition time between maximum current and zero current can be as short as a few tens of *picoseconds* (units of 10^{-12} seconds). The sudden transition—essentially a "vertical cliff" on a graph of current versus time—produces the harmonic output.

Hot-Carrier Diodes

Another specialty diode, called a *hot-carrier diode* (HCD), finds widespread application in RF circuits. It can function at much higher frequencies than can most other types of semiconductor diodes because it's "quiet," meaning that it generates very little *electrical noise* (chaotic output that covers a broad range of frequencies). This feature interests engineers who design wireless communications receivers.

The HCD has a high avalanche voltage. The junction capacitance is low, in part because the diode has a tiny point contact (Fig. 5-8). The wire is plated with gold or platinum to minimize corrosion. An N type silicon wafer serves as the semiconductor. The silicon wafer is coated with a thin film of aligned atoms called an *epitaxial layer*, produced by a lab process called *epitaxy*, in which technicians "grow" layers of semiconductor material on a base called a *substrate*. The epitaxial layer increases the avalanche voltage and reduces the junction capacitance, maximizing both the working voltage and the switching speed.

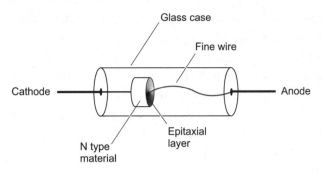

FIGURE 5-8 · Construction of a hot-carrier diode.

TIP *Epitaxy resembles the accumulation of frost on your car windows when the temperature falls below freezing and the humidity remains high. The manufacturers of HCDs "grow" the epitaxial layers under carefully controlled conditions.*

Signal Mixing

A *mixer* circuit combines two input signals having different frequencies to produce an output signal at either the sum or the difference of the input frequencies. The desired input signal (the one we want to "hear" in the output, such as a radio broadcast station) comes into the mixer from preceding amplifiers. The other input signal comes from a special circuit called a *local oscillator* (LO), whose frequency we can precisely control.

Consider two signals with frequencies f_1 and f_2, where f_1 represents the lower frequency and f_2 represents the higher frequency, with both frequencies expressed in the same units. If we combine these signals in a nonlinear device such as a diode, new signals result. One output signal occurs at the difference frequency (let's call it f_-), so we have

$$f_- = f_2 - f_1$$

The other output signal occurs at the sum frequency (let's call it f_+), giving us

$$f_+ = f_2 + f_1$$

We call the two output signals *mixing products*, and we call the frequencies f_- and f_+ *beat frequencies*.

Mixing is used extensively in wireless receivers and transmitters. The technique works quite well when we want to convert two input signals, one or both of which might range over a wide span of frequencies, to an output signal that stays at the same frequency all the time.

TIP   *Engineers find a signal that changes frequency only a little, or not at all, far easier to process and amplify than a signal whose frequency varies over a wide range. That's the main reason why signal mixers exist in RF communications equipment.*

Balanced Mixers

We can build a *single balanced mixer* (Fig. 5-9) cheaply and easily using HCDs. This circuit works at frequencies up to several gigahertz. The main disadvantage of this scheme lies in the fact that the *input port* and the *output port* aren't very well *isolated*, meaning that some of either input signal leaks through to the output. If we require excellent *input/output isolation*, we're better off using a *double balanced mixer*.

The input signals in a double balanced mixer (Fig. 5-10) do not leak through to the output nearly as much as they do in a single balanced mixer. In other words, little or no *coupling* (signal transfer or interaction) takes place among the three ports (signal input, LO input, and signal output). Hot-carrier diodes work well in this situation, just as they do in the single-balanced mixer. The diodes can handle large signal amplitudes without distortion, and they generate very little electrical noise.

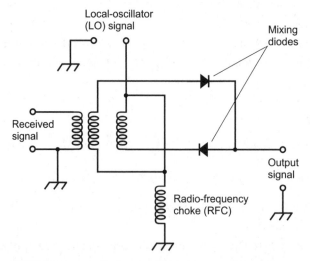

FIGURE 5-9 · A single balanced mixer using two hot-carrier diodes.

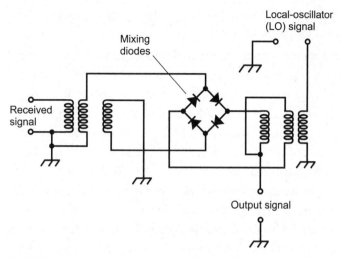

FIGURE 5-10 • A double balanced mixer using four hot-carrier diodes.

Both the single balanced mixer and the double balanced mixer have inherent *conversion loss*, meaning that the output signals are always weaker than either of the input signals. We can overcome conversion loss using one or more stages of amplification following the mixer.

Still Struggling

If we expect a diode to perform efficiently as a frequency multiplier or mixer, it must also work well as a detector at the same frequencies. The component should *not* behave as a capacitor at the signal input and output frequencies. In RF applications, PIN diodes, point-contact diodes, step-recovery diodes, and hot-carrier diodes can serve as frequency multipliers and mixers. Power-supply rectifier diodes, varactor diodes, and Zener diodes don't work well as frequency multipliers or mixers at RF because they have too much junction capacitance.

Amplitude Limiting

The forward breakover voltage of a germanium-based diode is about 0.3 V. For a diode made from silicon, it's about 0.6 V. A diode always conducts when the forward bias exceeds the breakover threshold, and the potential difference across the P-N junction remains essentially constant under these conditions: 0.3 V for germanium and 0.6 V for silicon.

Regardless of how much or how little current flows, the potential difference between the anode and the cathode varies little, or not at all, from the forward breakover voltage. We can use this property to advantage when we want to limit, or *clip*, the peak voltage of a signal. When we connect two identical diodes in *reverse parallel* (that is, in parallel with opposing polarities, as shown in Fig. 5-11A), the maximum positive or negative peak signal voltage cannot exceed the forward breakover voltage of the diodes. Figure 5-11B illustrates the input and output waveforms of a hypothetical *clipped* ("chopped off") signal that has passed through the circuit shown at A.

A diode limiter circuit always introduces signal distortion when clipping occurs. This phenomenon does not necessarily pose a problem with digital data, or with analog signals that rarely reach the limiting voltage. However, for analog voice or video signals with amplitude peaks that rise past the limiting voltage, the so-called *clipping distortion* degrades the fidelity. In the worst cases, the effect can render a voice signal unintelligible or mutilate an image signal beyond recognition.

TIP *Engineers in the "olden days of radio" (prior to about 1960) used simple diode clippers in the audio stages of communications receivers to prevent sudden, loud audio output bursts (a phenomenon called blasting) when a strong signal came in.*

Noise Limiting

In a radio communications receiver, a *noise limiter* can consist of a pair of diodes connected in reverse parallel with variable reverse bias for control of the clipping

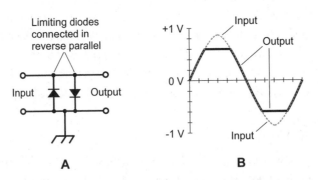

FIGURE 5-11 · At A, a limiter circuit using two diodes. At B, clipping of a signal waveform as it passes through a diode-based limiter circuit.

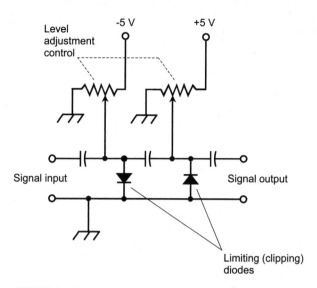

FIGURE 5-12 · A variable-threshold limiter circuit

level (Fig. 5-12). We can adjust the bias using a *dual potentiometer*, which comprises a pair of identical variable resistors connected to a single rotatable shaft.

A limiting circuit such as this one makes it possible to set the *clipping threshold voltage* from a minimum equal to the diode forward breakover voltage to a maximum equal to the forward breakover voltage plus the supply voltage (in this case ±5 V). When noise accompanies the desired signal, we can adjust the dual potentiometer until clipping occurs at the signal amplitude, so that noise pulses can never get stronger than the desired signal.

Frequency Control

When a diode is reverse-biased, the depletion region at the P-N junction has dielectric properties that make the diode act like a capacitor. The width of this zone depends on several factors, including the reverse-bias voltage. As long as the reverse bias remains less than the avalanche voltage, the width of the depletion region can be changed by varying the bias, causing a change in the junction capacitance. This capacitance, which rarely amounts to more than a few picofarads, varies *inversely* in proportion to the *square root* of the reverse bias voltage.

The varactor diode, mentioned earlier in this chapter, finds its "niche" in a circuit called a *voltage-controlled oscillator* (VCO). Figure 5-13 illustrates a parallel voltage-tuned *LC* circuit using a single varactor. As the reverse-bias voltage across the varactor fluctuates, so does the junction capacitance. This

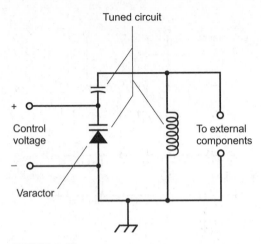

FIGURE 5-13 • A variable inductance-capacitance (*LC*) tuned circuit using a varactor.

phenomenon affects the resonant frequency of the *LC* circuit, and therefore, the frequency at which the oscillator generates its output. The fixed capacitor, whose value should greatly exceed the varactor capacitance, prevents the inductor coil from "shorting out" the control-voltage source.

TIP   *The schematic symbol for a varactor diode has two lines, rather than one, on the cathode side, emphasizing the fact that the capacitive effect of the depletion region is utilized, rather than the "one-way-current" property of the P-N junction.*

 PROBLEM 5-3

Suppose that we apply two signals, one having a frequency of 3.500 MHz and the other having a frequency of 12.500 MHz, to the input of a signal mixer. What output signal frequencies will we get, not including any input that "leaks through"?

✔**SOLUTION**

The output frequencies equal the sum and difference of the input signal frequencies. In this example, let $f_1 = 3.500$ MHz and $f_2 = 12.500$ MHz. Then we have

$$f_- = f_2 - f_1$$
$$= 12.500 - 3.500$$
$$= 9.000 \text{ MHz}$$

and

$$f_+ = f_2 + f_1$$
$$= 12.500 + 3.500$$
$$= 16.000 \text{ MHz}$$

PROBLEM 5-4

Imagine that we want to design a mixer circuit that will allow the reception of radio signals over a continuous span from 3.500 MHz to 4.000 MHz. What range of frequencies should the LO have if we want to generate a difference-frequency signal at a constant 9.000 MHz?

SOLUTION

We must obtain input signals whose frequency difference, $f_2 - f_1$, always equals 9.000. Let f_1 represent the frequency of the signal that we want to "hear." It can range from 3.500 MHz to 4.000 MHz. The LO frequency f_2 must always exceed f_1 by 9.000 MHz if we want to get a difference output at 9.000 MHz. The tunable range of the LO should therefore run from a minimum of 9.000 + 3.500 = 12.500 MHz to a maximum of 9.000 + 4.000 = 13.000 MHz.

Oscillation and Amplification

Under some conditions, diodes can generate or amplify radio signals at ultra-high frequencies (UHF) and microwave radio frequencies. All frequencies in this part of the RF spectrum exceed 300 MHz.

Gunn Diodes

A *Gunn diode* can produce up to 1 W of RF power output, but usually it works at levels of only about 0.1 W. Most Gunn diodes are made from gallium arsenide. The device produces signals because of the so-called *Gunn effect*, named after J. Gunn of International Business Machines (IBM), who observed the phenomenon in the 1960s and exploited it to construct UHF and microwave oscillators.

Gunn-diode oscillation takes place as a result of a property called *negative resistance*. This term refers to the fact that within a certain region of the diode's characteristic curve, the current decreases as the voltage increases. Negative resistance produces electrical instability leading to the generation of RF signals. Under certain conditions, we can use a varactor to adjust the output frequency of a Gunn-diode oscillator.

TIP   *A Gunn-diode oscillator, connected directly to a microwave* **horn antenna,** *is known as a* **Gunnplexer.** *These devices enjoy considerable popularity among electronics experimenters, especially amateur ("ham") radio operators.*

IMPATT Diodes

The acronym *IMPATT* (pronounced "IM-pat") comes from the words *impact avalanche transit time*, describing an electrical effect that resembles negative resistance. An IMPATT diode is a microwave oscillating device like a Gunn diode, except that it's made from silicon rather than gallium arsenide. An IMPATT diode can operate as an amplifier for a microwave transmitter that employs a Gunn-diode oscillator. As an oscillator, an IMPATT diode produces about the same output power, at comparable frequencies, as a Gunn diode does.

Tunnel Diodes

Another type of diode that can oscillate at microwave frequencies goes by the interesting name *tunnel diode*. Some engineers call it an *Esaki diode*. It produces only a tiny amount of RF power output, but nevertheless enough to let the device serve as the LO in a microwave radio receiver. Tunnel diodes, especially those made from GaAs, work well as amplifiers in microwave receivers because they generate minimal noise. At UHF and microwave frequencies, engineers constantly strive to keep internally generated circuit noise to a minimum.

Diodes and Radiant Energy

Some semiconductor diodes emit *infrared* (IR) or visible-light energy when current passes through the P-N junction in a forward direction. This phenomenon, called *photoemission*, occurs as electrons "fall" from relatively higher to relatively lower energy states within atoms. Diodes can respond to incoming IR or visible light as well. Some *photosensitive diodes* have variable resistance that depends on the illumination intensity. Others generate their own DC voltages in the presence of IR or visible light.

LEDs and IREDs

The most common color for a *light-emitting diode* (LED) was originally bright red. Nowadays, we can find LEDs that produce output in practically any color. An *infrared emitting diode* (IRED) produces wavelengths too long for our eyes to see, just "below" the visible red end of the spectrum. The intensity of the energy emission from an LED or IRED depends on the forward current. As the current rises, the brightness increases up to a certain point. If the current continues to rise, no further increase in brilliance takes place. Then we say that the device operates in a state of *saturation*.

Laser Diodes

The *laser diode* is a specialized LED or IRED with a relatively large, flat P-N junction, built on top of a metal or silicon base called a *substrate*. The junction emits *coherent radiation* if we drive sufficient forward current through it. In coherent radiation, all of the EM energy occurs at a single wavelength, and all of the wavefronts "line up" in phase with each other.

If the forward current remains below a certain threshold, a laser diode behaves like an ordinary LED or IRED. When the current reaches or exceeds the critical level, the charge carriers near the P-N junction recombine so that *lasing* (the effect that produces coherent radiation) takes place. Figure 5-14 shows the construction of a typical laser diode, also known as an *injection laser*.

Laser diodes work especially well in applications where we want visible or IR rays to travel a long way through clear air or free space. As the rays propagate

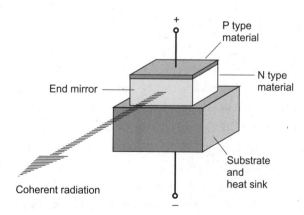

FIGURE 5-14 • Construction of a laser diode.

over great distances, coherent light or IR suffers less attenuation than ordinary light or IR does because we can focus the rays into a narrow, "parallel" energy beam that doesn't spread out very much as it travels away from the source.

Communications

Visible-light and IR-emitting laser diodes find applications in communications systems because we can *modulate* (vary) the radiation intensity to carry information. When the current through the device rises to a level sufficient to produce radiant energy but not enough to cause saturation, the instantaneous visible or IR output follows along with rapid current changes. This allows us to "imprint" voice, music, video, or digital data onto visible or IR rays.

TIP *Many data communications systems, especially in high-speed computer networks, make use of modulated visible-light or IR transmitted through clear fibers, an application known as* **fiber-optic technology.**

Silicon Photodiodes

A silicon diode, housed in a transparent case and constructed so that visible light can strike the P-N junction, forms a *photodiode*. We apply a reverse DC bias to the device. When visible light, IR, or ultraviolet (UV) rays strike the junction, current can flow easily through it. The current varies in proportion to the intensity of the incoming radiation, within certain limits. When energy of varying intensity strikes the P-N junction of a reverse-biased silicon photodiode, the current through the device follows right along with the fluctuations in incident illumination. This effect makes silicon photodiodes useful for receiving modulated-light, modulated-IR, or modulated-UV signals.

Optoisolators

An LED or IRED and a photodiode can be housed in a single package to obtain an *optoisolator*, also known as an *optical coupler* or *optocoupler*. With an electrical-signal input, the LED or IRED generates a modulated visible-light or IR beam and sends it over a small, clear gap to the photodiode, which converts the radiant energy back into an electrical signal.

Whenever we electrically couple one electronic circuit to another, we should expect that the two circuits will interact. For example, an oscillator might work just fine all by itself, but when we connect it to an amplifier in an attempt to get a stronger signal, the oscillator can become unstable because of the *load* that the amplifier imposes on it. An optoisolator between the oscillator and the amplifier can eliminate this nuisance.

Still Struggling

So-called *impedance interaction* among electronic circuits can't take place over visible-light or IR beams. If the electrical input impedance of the second circuit varies, even drastically, the impedance that the first circuit "sees" doesn't change at all. The first circuit "sees" only the load caused by the LED or IRED inside the optoisolator. This load remains constant no matter what happens on the photo-diode side.

Photovoltaic Cells

A silicon diode, with no bias voltage applied, generates DC all by itself if sufficient visible light, IR, or UV energy strikes its P-N junction. Physicists and engineers call this phenomenon the *photovoltaic effect*. It's the principle by which *solar cells* and *solar panels* work.

A typical *photovoltaic (PV) cell* has a P-N junction with a large surface area (Fig. 5-15), maximizing the amount of radiant energy that can strike the junction.

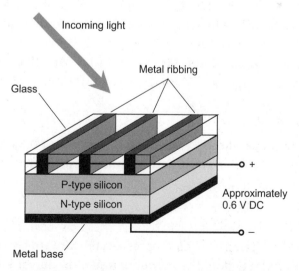

FIGURE 5-15 · Construction of a photovoltaic (PV) cell.

A single silicon PV cell produces approximately 0.6 V DC in direct sunlight with no load. The amount of current that it can deliver, given a certain amount of incident radiant power, depends on the surface area of the junction.

Semiconductor-device vendors routinely connect multiple silicon PV cells in *series-parallel* arrays (that is, sets of series-connected cells in parallel, or sets of parallel-connected cells in series) to provide solar power for small electronic appliances, such as portable radios and personal computers. A large assembly of PV cells constitutes a *solar panel*.

In a theoretically ideal situation where all PV cells have no internal resistance, we can state four "rules" concerning series or parallel combinations of PV cells subjected to constant incident radiation.

1. The maximum obtainable voltage (MOV) across two or more identical series-connected PV cells increases in direct proportion to the number of cells in the set.

2. The MOV across two or more identical parallel-connected PV cells equals the MOV of a single cell.

3. The maximum deliverable current (MDC) from two or more identical parallel-connected PV cells increases in direct proportion to the number of cells in the set.

4. The MDC from two or more identical series-connected PV cells equals the MDC from a single cell.

In practical scenarios, these rules hold up fairly well, as long as we don't combine more than a few PV cells in series within an array. If we connect too many cells in series, their internal resistances (which, in reality, don't exactly equal zero) will add up to something significant, reducing the MDC from the combination compared with the MDC that we would get from a single cell.

 PROBLEM 5-5

Why does a single silicon PV cell produce only about 0.6 V DC, no matter how much light we shine on it, and regardless of the surface area of the P-N junction?

SOLUTION

The potential difference of 0.6 V represents the forward breakover voltage of a silicon P-N junction. The output of a semiconductor PV cell appears with a polarity corresponding to forward bias. This phenomenon "internally limits" the voltage that can exist across the junction, in a manner similar to the way a diode-based limiter works.

PROBLEM 5-6

Suppose that a single silicon PV cell can produce, at the absolute maximum, 100 mA of current in direct sunlight. How many of these devices must we connect in parallel to obtain a maximum deliverable current of 500 mA?

SOLUTION

Five PV cells in parallel would serve this purpose. The maximum deliverable current values add up directly when we connect two or more PV cells in parallel.

PROBLEM 5-7

Imagine that we connect 16 identical silicon PV cells in a 4 × 4 series-parallel array, where each cell can provide 75 mA of current at most. We can create such an array in either of two ways, as follows:

- Construct four series-connected sets of four cells each, and then connect them all in parallel.
- Construct four parallel-connected sets of four cells each, and then connect them all in series.

What's the maximum obtainable voltage (MOV) that we can expect from this array? What's the maximum deliverable current (MDC)?

SOLUTION

Carefully read the four "rules" for series and parallel combinations of PV cells, stated above. The MOV equals 0.6 V × 4, or 2.4 V. The MDC equals 75 mA × 4, or 300 mA.

QUIZ

Refer to the text in this chapter if necessary. A good score is eight correct. The answers are listed in the back of the book.

1. In a PIN diode, the intrinsic semiconductor layer serves primarily to
   A. increase the forward breakover voltage.
   B. decrease the avalanche voltage.
   C. increase the current-handling capability.
   D. decrease the junction capacitance.

2. Which of the following devices would work best for intercepting modulated light rays?
   A. An LED
   B. A laser diode
   C. A silicon photodiode
   D. A varactor diode

3. Figure 5-16 illustrates a single balanced mixer circuit. What diode type should we choose for this application?
   A. Zener
   B. Hot-carrier
   C. Rectifier
   D. Varactor

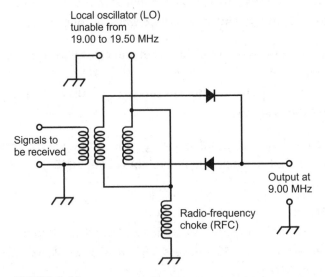

FIGURE 5-16 · Illustration for Quiz Questions 3 and 4.

4. Which of the following received-signal frequency ranges can we "hear" using the mixer circuit shown in Fig. 5-16?

    A. 10.00 to 10.50 MHz
    B. 8.50 to 9.00 MHz
    C. 10.00 to 28.50 MHz
    D. 14.00 to 14.25 MHz

5. A double balanced mixer using four diodes has at least one outstanding advantage over a single balanced mixer using two diodes. What's that advantage?

    A. Wider operating frequency range
    B. Better isolation among the signal ports
    C. Greater gain
    D. Less tendency to oscillate

6. In a "crystal-set" AM radio receiver, a point-contact diode can effectively serve as the

    A. local oscillator.
    B. audio clipper.
    C. envelope detector.
    D. frequency multiplier.

7. Photoemission in a semiconductor material occurs as a result of

    A. electrons "falling" from higher to lower energy states.
    B. electrons combining with holes.
    C. the sudden disappearance of a depletion region.
    D. avalanche effect.

8. Suppose that we apply a small reverse-bias voltage to a rectifier diode, well below the avalanche threshold. If we increase the reverse-bias voltage gradually but indefinitely, what effects do we observe?

    A. The junction capacitance gradually increases, and then, at a certain critical voltage, the diode acts as an open circuit.
    B. The junction capacitance gradually increases, and then, at a certain critical voltage, the diode acts as a short circuit.
    C. The junction capacitance gradually decreases, and then, at a certain critical voltage, the diode acts as an open circuit.
    D. The junction capacitance gradually decreases, and then, at a certain critical voltage, the diode acts as a short circuit.

9. Imagine that we find a Zener diode in our "spare component bin" and connect that diode in series with a variable-voltage source and a current-limiting resistor. Then we apply various forward-bias and reverse-bias voltages to the diode

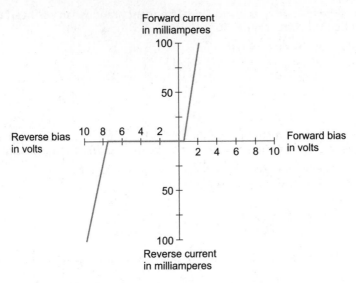

FIGURE 5-17 · Illustration for Quiz Questions 9 and 10.

while measuring the current through it. We tally up all of our current-versus-voltage data and plot it as a graph, getting Fig. 5-17. Approximately what Zener voltage does this device appear to have?

A. 0.6 V

B. 4.1 V

C. 7.6 V

D. 8.1 V

10. The Zener voltage in the graph of Fig. 5-17 coincides with the

A. forward breakover voltage.

B. avalanche voltage.

C. average of the forward breakover and avalanche voltages.

D. sum of the forward breakover and avalanche voltages.

chapter 6

Transistors and Integrated Circuits

Transistors can generate signals, change weak signals into strong ones, mix signals, and act as high-speed switches. *Integrated circuits* (ICs) contain many transistors, along with other components, on a thin, flat piece of semiconductor material called a *chip*.

CHAPTER OBJECTIVES

In this chapter you will

- See how bipolar transistors behave as the voltages and currents vary.
- Learn how bipolar transistors amplify and oscillate.
- See how field-effect transistors behave as the voltages and currents vary.
- Learn how field-effect transistors amplify and oscillate.
- Discover what happens as you vary the AC signal frequency in a transistor.
- Familiarize yourself with the basic types and applications of ICs.

Bipolar Transistors

Bipolar transistors have three sections of semiconductor material with two P-N junctions. Two major geometries exist: a P type layer between two N type layers (called an *NPN transistor*), or an N type layer between two P type layers (called a *PNP transistor*).

NPN versus PNP

Figure 6-1A is a simplified functional drawing of an NPN transistor, and Fig. 6-1B shows the symbol that engineers use to represent it in schematic diagrams. The P type, or center, layer constitutes the *base*. One of the N type layers forms the *emitter*, and the other N type layer forms the *collector*. We can label the base, the emitter, and the collector as B, E, and C, respectively.

A PNP transistor has two P type layers, one on either side of a thin N type layer as shown in Fig. 6-1C. The schematic symbol for this device appears in Fig. 6-1D. The N type layer constitutes the base. One of the P type layers forms the emitter, and the other P type layer forms the collector. As with the NPN device, we can label the three electrodes as B, E, and C.

We can easily tell from a schematic diagram whether the circuit designer means for a particular transistor to be NPN type or PNP type. Once we realize

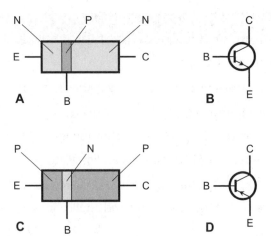

FIGURE 6-1 · Pictorial diagram of an NPN transistor (A), the schematic symbol for an NPN transistor (B), pictorial diagram of a PNP transistor (C), and the schematic symbol for a PNP transistor (D).

that the arrow always goes with the emitter, we can identify the three electrodes without having to label them. In an NPN transistor, the arrow at the emitter points outward. In a PNP transistor, the arrow at the emitter points inward.

In electronic circuits, PNP and NPN transistors can perform similar functions. However, they require the opposite voltage polarities, and the currents flow in the opposite directions. In many situations, we can replace an NPN device with a PNP device or vice versa, reverse the power-supply polarity, and expect the new circuit to work the same with the replacement device, provided that the replacement transistor has the appropriate specifications.

TIP *Some bipolar transistors function primarily in radio-frequency (RF) amplifiers and oscillators; others work better for audio-frequency (AF) work. Some can handle high power, and others are made for weak-signal applications. Some are manufactured for digital switching, and others work well for analog signal processing.*

Power Supply for an NPN Device

In order to make any transistor work, we must apply certain voltages to it, called *bias voltages* (or simply *bias*). In an NPN transistor, we usually connect the emitter to the negative battery or power-supply terminal and the collector to the positive terminal, often through coils and/or resistors. Figure 6-2 shows the basic biasing scheme for an NPN transistor. Typical battery or power-supply voltages range from 3 V to 50 V.

In Fig. 6-2, we label the base electrode "Control" because, with constant collector-to-emitter (C-E) voltage (symbolized E_C or V_C), the magnitude of the current through the transistor depends on the current that flows across the

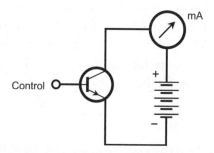

FIGURE 6-2 · Power supply connection to the emitter and collector of an NPN transistor.

emitter-base (E-B) junction, called the *base current* and labeled I_B. What happens at the base, from instant to instant in time, dictates how the entire device behaves.

Zero Bias

When we don't connect the base to anything, or when we short-circuit it to the emitter for DC, we say that a bipolar transistor operates at zero *base bias* (or simply at *zero bias*). No appreciable current can flow between the emitter and the collector with zero bias under no-signal conditions. In order to cause current to flow between the emitter and collector, we must bias the emitter-base (E-B) junction as we would do with a semiconductor diode to make it conduct in the forward direction. We must either apply a DC voltage at least equal to the *forward breakover* voltage at the E-B junction, or else introduce an AC signal at the base or emitter that causes the voltage at the E-B junction to reach or exceed forward breakover for at least a small part of each signal cycle. In a silicon transistor, this minimum forward voltage equals approximately 0.6 V. For germanium transistors, it's close to 0.3 V.

Reverse Bias

In the scenario of Fig. 6-2, imagine that we connect a second battery between the base and the emitter of the NPN transistor, forcing the base to acquire a negative voltage with respect to the emitter. The addition of this new battery will cause the E-B junction to operate in a condition of *reverse bias*. Let's assume that this new battery doesn't have enough voltage to cause avalanche breakdown at the E-B junction.

When we reverse-bias the E-B junction of a transistor, no current flows between the emitter and the collector under no-signal conditions. We might inject a signal at the base or emitter to overcome the combined reverse-bias battery voltage and forward-breakover voltage of the E-B junction, but such a signal must have positive voltage peaks high enough to cause conduction at the E-B junction for part of the input signal cycle. Otherwise, the transistor will remain in a state called *cutoff* for the entire cycle.

Forward Bias

Now suppose that, in the situation of Fig. 6-2, we make the bias voltage at the base of the NPN transistor positive relative to the emitter, starting at small levels and gradually increasing. We might do this by connecting a fairly large-value resistor between the base and the existing positive battery terminal. Alternatively, we can do it with a second battery. This action causes forward bias at the E-B junction. If this bias remains smaller than forward breakover, no current flows. But once the DC voltage reaches and then exceeds the forward breakover threshold, the E-B junction conducts under no-signal conditions.

Despite reverse bias at the base-collector (B-C) junction, some *emitter-collector current*, more often simply called *collector current* and denoted I_C, flows when the E-B junction conducts. A small instantaneous rise in the *positive* polarity of an AC signal at the base, attended by a small instantaneous rise in the base current I_B, will cause a large instantaneous increase in the collector current I_C. Conversely, a small instantaneous rise in the *negative* polarity of an AC signal at the base, attended by a small instantaneous drop in I_B, will cause a large instantaneous decrease in I_C. Therefore, the collector current varies a lot more than the base current does, and the transistor acts as an AC *current amplifier*.

Saturation

If we adjust the conditions at the E-B junction so that I_B continues to rise, we'll eventually reach a point where I_C no longer increases rapidly. Ultimately, the I_C versus I_B function, or *characteristic curve*, levels off. Figure 6-3 shows a *family of characteristic curves* for a hypothetical bipolar transistor. Each individual curve in the set portrays the situation for a certain fixed *collector-to-emitter voltage,*

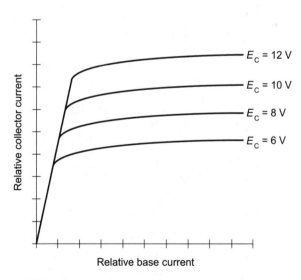

FIGURE 6-3 · A family of characteristic curves for a bipolar transistor.

usually called simply the *collector voltage* and symbolized as E_C. The actual current levels depend on the internal structure of the transistor. The overall current levels tend to be large for power transistors and small for weak-signal transistors. (That's why we don't label the graph axes with numbers in Fig. 6-3.) Where the curves level off, we say that the transistor operates in a state of *saturation*. Under these conditions, it won't work as an amplifier.

TIP   *When working with analog circuits such as AC signal amplifiers, engineers rarely, if ever, want to operate bipolar transistors in the saturated state. However, engineers often bias bipolar transistors at saturation in digital circuits where the signal is always either "all the way on" (high, or logic 1) or "all the way off" (low, or logic 0).*

Power Supply for a PNP Device

For a PNP transistor, the DC power supply connection to the emitter and collector constitutes a "mirror image" of the case for an NPN device, as shown in Fig. 6-4. We reverse the battery polarity compared with the NPN circuit. To overcome forward breakover at the E-B junction, an applied voltage or signal at the base ("Control" in the figure) must attain sufficient negative peak polarity. If we apply a positive DC voltage at the base, the device will remain in a state of cutoff.

Either type of transistor—PNP or the NPN—can serve as a *current valve*. Small changes in the base current I_B induce large fluctuations in the collector current I_C when we operate the device in the region of the characteristic curve where the graph has a steep "rise over run" (technically called *slope*). The internal

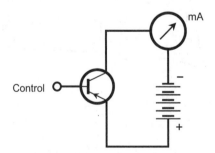

FIGURE 6-4 · Power supply connection to the emitter and collector of a PNP transistor.

atomic activity differs in the PNP device as compared with the NPN device, but in most practical cases, the external circuitry "can't tell the difference."

Static Current Characteristics

We can describe the current-carrying characteristics of a bipolar transistor in simplistic terms as the *static forward current transfer ratio*. This parameter comes in two "flavors:" one (symbolized H_{FB}) that describes the collector current versus the emitter current when we place the base at electrical ground, and the other (symbolized H_{FE}) that describes the collector current versus the base current when we place the emitter at electrical ground.

Mathematically, the quantity H_{FB} equals the ratio of the collector current to the emitter current at any given moment in time with the base grounded:

$$H_{FB} = I_C/I_E$$

For example, if an emitter current I_E of 100 mA results in a collector current I_C of 90 mA, then we can calculate

$$H_{FB} = 90/100$$
$$= 0.90$$

If $I_E = 100$ mA and $I_C = 95$ mA, then

$$H_{FB} = 95/100$$
$$= 0.95$$

The quantity H_{FE} equals the ratio of the collector current to the base current at any given moment in time with the emitter grounded:

$$H_{FE} = I_C/I_B$$

For example, if a base current I_B of 10 mA results in a collector current I_C of 90 mA, then we can calculate

$$H_{FE} = 90/10$$
$$= 9.0$$

If $I_B = 5.0$ mA and $I_C = 95$ mA, then

$$H_{FE} = 95/5.0$$
$$= 19$$

Alpha

We can describe the current variations in a bipolar transistor by dividing the *difference* in I_C by the *difference* in I_E that occurs when we apply a small signal to the emitter of a transistor with the base connected to electrical ground (or placed at the same potential difference as electrical ground). We call this ratio the *alpha*, symbolized by the italic lowercase Greek letter alpha (α). Let's abbreviate the words "the difference in" by writing an uppercase nonitalic Greek letter delta (Δ). Then mathematically, we can define

$$\alpha = \Delta I_C / \Delta I_E$$

Engineers and technicians call this quantity the *dynamic current gain* of the transistor in the grounded-base situation. In this context, the term "gain" means "amplification factor." The alpha for any transistor is always less than 1 in a circuit where collector current flows for at least part of the AC input signal cycle. That's because whenever we apply a signal to the input, the base "bleeds off" at least a little bit of current from the emitter before it shows up at the collector. Technically, the alpha constitutes a *current loss* (a negative gain).

TIP   *In a cut-off transistor where the AC input signal remains too weak to produce collector current during any part of the cycle, we can't technically define the alpha because $\Delta I_C = 0$ and $\Delta I_E = 0$.*

Beta

We get an excellent definition of current amplification for "real-world signals" when we divide the difference in I_C by the difference in I_B that occurs when we apply a small signal to the base of a transistor with the emitter at electrical ground. Then we get the dynamic current gain for the grounded-emitter case. We call this ratio the *beta*, symbolized as an italic lowercase Greek letter beta (β). Once again, let's abbreviate the words "the difference in" as Δ. Then we have

$$\beta = \Delta I_C / \Delta I_B$$

The beta for a transistor can exceed 1. Under some conditions it can attain values into the hundreds. Under other conditions, we might observe a beta of less than 1. We might get a beta smaller than 1 if we improperly bias a transistor, if we choose the wrong type of transistor for a particular application, or if we try to operate the transistor at a signal frequency that's too high.

TIP *In a cut-off transistor where the AC input signal never gets strong enough to produce collector current during any part of the cycle, the beta equals 0 because $\Delta I_C = 0$ while ΔI_B has some finite (although tiny) "leakage" value.*

How Alpha and Beta Relate

Whenever nonzero base current flows in a bipolar transistor, we can calculate the beta in terms of the alpha with the formula

$$\beta = \alpha / (1 - \alpha)$$

and we can calculate the alpha in terms of the beta with the formula

$$\alpha = \beta / (1 + \beta)$$

If you're a mathematician, you can derive both of these formulas from the fact that, at any instant in time, the collector current equals the emitter current minus the base current. That is,

$$I_C = I_E - I_B$$

Alpha Cutoff Frequency

Suppose that we operate a bipolar transistor as a current amplifier, and we deliver an input signal to it at a frequency of 1 kHz. Then we steadily increase the input-signal frequency. We notice that the value of α declines as the frequency goes up. We define the *alpha cutoff frequency* of a bipolar transistor, symbolized f_α, as the frequency at which α decreases to 0.707 times its value at 1 kHz. (Don't confuse this use of the term *cutoff* with the state of cutoff that we get when we zero-bias or reverse-bias a transistor under no-signal conditions!) A transistor can have considerable dynamic current gain (beta) at, and even well above, its alpha cutoff frequency. By looking at this specification for a particular transistor, we can get an idea of how rapidly it loses its ability to amplify as the frequency goes up.

Beta Cutoff Frequency

Imagine that we repeat the above-described variable-frequency experiment. We'll discover that the value of β goes down as the frequency increases. We define the *beta cutoff frequency* (also called the *gain bandwidth product*) for a bipolar transistor, symbolized f_β or f_T, as the frequency at which the value of β

declines all the way down to 1. If we try to make an amplifier using a transistor at a frequency higher than its beta cutoff specification, we will never succeed! A transistor *cannot* exhibit any dynamic current gain above its beta cutoff frequency.

PROBLEM 6-1

What's the static forward current transfer ratio H_{FE} in a device where a base current of 3.0 mA produces a collector current of 24 mA?

SOLUTION

Use the formula for H_{FE}, as follows:

$$H_{FE} = I_C / I_B$$

In this case, the base current, I_B, equals 3.0 mA and the collector current, I_C, equals 24 mA. Therefore

$$H_{FE} = 24 / 3.0$$
$$= 8.0$$

PROBLEM 6-2

What's the beta in a transistor where a change in base current of 0.10 µA results in a change in the collector current of 1.0 µA?

SOLUTION

Remember the formula for beta in terms of a change in the collector current that results from a change in the base current:

$$\beta = \Delta I_C / \Delta I_B$$

In this case, ΔI_C = 1.0 µA and ΔI_B = 0.10 µA. Therefore:

$$\beta = 1.0 / 0.10$$
$$= 10$$

Basic Bipolar-Transistor Circuits

We can connect a bipolar transistor to external components to build electronic circuits that do specialized tasks, such as amplification and oscillation. Three general arrangements exist:

1. The *common-emitter circuit*, in which we ground the emitter for AC signals
2. The *common-base circuit*, in which we ground the base for AC signals
3. The *common-collector circuit*, in which we ground the collector for AC signals

Common Emitter

Figure 6-5 is a schematic diagram of a generic NPN common-emitter circuit. Capacitor C_1 presents a short circuit to the AC signal, placing the emitter at *signal ground*. Resistor R_1 gives the emitter a small positive DC voltage with respect to ground. The exact DC voltage at the emitter under no-signal conditions depends on the value of R_1 and also on the base bias, determined by the ratio of the values of resistors R_2 and R_3. The base bias can range from zero all the way up to the supply voltage, in this case +12 V DC. Normally, the base bias is a couple of volts positive with respect to DC ground.

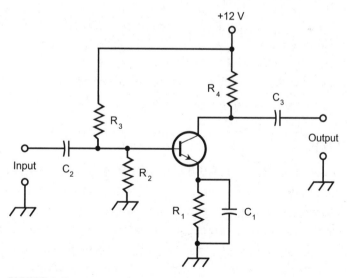

FIGURE 6-5 · Common-emitter NPN configuration.

Capacitor C_2 isolates, or *blocks*, DC from the input while allowing the AC signal to get through with ease. Capacitor C_3 blocks DC from the output while letting the AC signal pass. Engineers and technicians call capacitors such as C_2 and C_3 *blocking capacitors*. Resistor R_4 keeps the output signal from shorting through the power supply while allowing a positive DC voltage to exist at the collector. A signal enters the circuit through C_2, where it causes the base current, I_B, to vary. Small fluctuations in I_B cause large variations in I_C. This current passes through R_4, causing a fluctuating DC voltage to appear across that resistor. The AC signal, superimposed on the DC flowing through R_4, passes through C_3 to the output terminal.

The common-emitter configuration offers excellent gain when properly designed to operate as an amplifier. The AC output wave appears in phase opposition to the input wave; if we input a sine wave to the circuit, this inversion constitutes the equivalent of a 180° phase shift. If circuitry outside the transistor inverts the signal again, a high-gain common-emitter circuit can generate a signal of its own. Then we say that it *breaks into oscillation*. Under these conditions, the circuit can, and usually will, fail to function as a good amplifier.

Still Struggling

We can minimize the risk of oscillation in a common-emitter amplifier by making sure that the circuit doesn't have too much gain. (Believe it or not, an amplifier can sometimes "work too well.") Alternatively, we can use a different circuit configuration, such as the common-base design described next.

Common Base

In the common-base circuit (Fig. 6-6), we place the base at signal ground. The DC bias on the transistor is the same for this circuit as for the common-emitter circuit. We apply the AC input signal to the emitter, producing fluctuations in the voltage across R_1, in turn causing small variations in I_B. As a result, we get a large change in I_C, the collector current that flows through R_4, so amplification occurs. The output wave appears in phase with the input wave.

The signal enters through C_1. Resistor R_1 keeps the input signal from shorting to ground. Base bias is provided by R_2 and R_3. Capacitor C_2 keeps the base

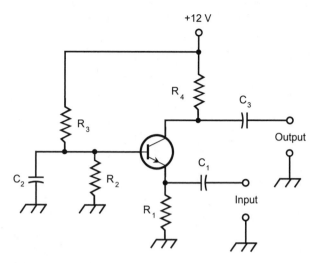

FIGURE 6-6 · Common-base NPN configuration.

at signal ground. Resistor R_4 keeps the signal from shorting through the power supply. The output signal goes through C_3.

A common-base circuit can't produce as much gain as a common-emitter circuit does, but the common-base amplifier is less prone to oscillate because it's less susceptible to the undesirable effects of *positive feedback*, which can cause an amplifier to "rage out of control" in much the same way that public-address system "howls" when the microphone picks up too much sound from the speakers.

TIP *Common-base circuits work well as high-power amplifiers in radio transmitters, where too much positive feedback can cause* **parasitic oscillation**—*so called because, like an "electronic parasite," it robs the transmitter of useful signal output power on its design frequency. Parasitic signals can also wreak havoc on the airwaves by interfering with other wireless communications systems.*

Common Collector

A common-collector circuit (Fig. 6-7) operates with the collector at signal ground. The input signal passes through C_2 onto the base of the transistor. Resistors R_2 and R_3 provide the base bias. Resistor R_4 limits the current through the transistor. Capacitor C_3 keeps the collector at signal ground. A fluctuating current flows through R_1, and a fluctuating voltage, therefore, appears across it. The AC component passes through C_1 to the output. Because the instantaneous output signal level follows along with the instantaneous emitter current, this

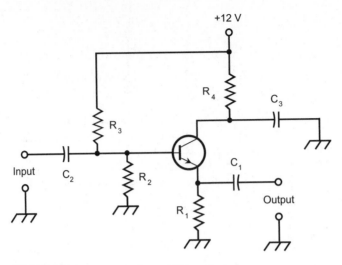

FIGURE 6-7 · Common-collector NPN configuration.

circuit is sometimes called an *emitter follower*. The output wave appears in phase with the input wave. When well designed, an emitter follower works over a wide range of frequencies, and offers a low-cost alternative to an RF transformer.

TIP *An emitter-follower circuit won't amplify signals, but it can help to provide isolation between two different parts of an electronic system. We call this sort of isolation circuit a* **buffer**.

Still Struggling

Figures 6-5, 6-6, and 6-7 show NPN transistor circuits. We can obtain the equivalent diagrams for PNP circuits by replacing the NPN transistors with PNP devices, and by reversing the power-supply polarity in each case (providing about −12 V DC, rather than +12 V DC, at the noncollector end of resistor R_4).

Field-Effect Transistors

The other major type of semiconductor transistor, besides the bipolar device, is the *field-effect transistor* (FET). Two main versions exist: the *junction FET* (JFET) and the *metal-oxide-semiconductor FET* (MOSFET).

Principle of the JFET

In a JFET, a fluctuating electric field causes the current to vary within the semi-conductor medium. Electrons or holes move along a path called the *channel* from the *source* (S) electrode to the *drain* (D) electrode. As a result, we get a drain current I_D equal to the source current I_S, and also equal to the current at any point along the channel. The current through the channel depends on the instantaneous voltage at the *gate* (G) electrode.

If we design a JFET circuit properly, small changes in the gate voltage E_G cause large changes in the current through the channel, and therefore, in I_D. When the fluctuating drain current passes through an external resistance, we get large variations in the instantaneous DC voltage across that resistance. We can "draw off" the AC part of the fluctuating DC, thereby obtaining an output signal that's much stronger than the input signal. That's how an FET produces *voltage amplification*.

N-Channel versus P-Channel

Figure 6-8A is a simplified pictorial drawing of an *N-channel JFET*. Figure 6-8B shows its schematic symbol. The N type material forms the path for the current. Electrons constitute most of the charge carriers; engineers would say that electrons are the *majority carriers*. The drain is connected to the positive power-supply terminal, often through a resistor, a coil, or some other combination of components. The gate comprises P type material. Another, larger section of P type material, called the *substrate*, forms a boundary on the side of the channel opposite the gate. The voltage on the gate produces an electric field that interferes with the flow of charge carriers through the channel. As E_G becomes more negative, the electric field chokes off the current though the channel to an increasing extent, so the drain current I_D decreases.

A *P-channel JFET* (Figs. 6-8C and D) has a current pathway made of P type semiconductor material. The majority carriers are holes. The drain is connected to the negative power-supply terminal. The gate and substrate consist of N type material. The more positive E_G gets, the more the electric field chokes off the current through the channel, and the smaller I_D becomes.

You can usually recognize an N-channel JFET in schematic diagrams by the presence of an arrow pointing inward at the gate, and a P-channel JFET by the presence of an arrow pointing outward at the gate. Some diagrams don't show these arrows, but the power-supply polarity gives away the device type. When the drain goes to the positive power-supply voltage (with the negative

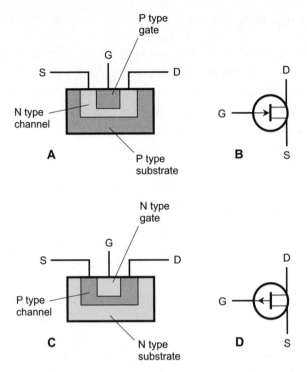

FIGURE 6-8 · Pictorial diagram of an N-channel JFET (A), the schematic symbol for an N-channel JFET (B), pictorial diagram of a P-channel JFET (C), and the schematic symbol for a P-channel JFET (D).

power-supply terminal connected to ground), it indicates an N-channel JFET. When the drain goes to the negative power-supply voltage (with the positive power-supply terminal usually connected to ground), it indicates a P-channel JFET.

TIP   *If we replace an N-channel JFET with a P-channel JFET and reverse the power-supply polarity, the new circuit will work the same as the old one did, as long as the new JFET has the proper specifications.*

Depletion and Pinchoff

A JFET works because the voltage at the gate produces an electric field that interferes, more or less, with the flow of charge carriers along the channel.

As the drain voltage E_D increases, so does the drain current I_D, up to a certain maximum "leveling-off" value—as long as the gate voltage E_G remains constant, and as long as E_G isn't too large. As E_G increases (negatively in an N channel or positively in a P channel), a *depletion region* forms in the channel. Charge carriers can't flow in the depletion region, so they must pass through a narrowed channel. Because of the restricted pathway, the current goes down.

As the gate voltage E_G increases (negatively for an N-channel device or positively for a P-channel device), the depletion region widens and the channel narrows. If E_G gets high enough, the depletion region closes off the channel altogether, preventing any flow of charge carriers from the source to the drain. We call this state of affairs *pinchoff*.

JFET Biasing

Figure 6-9 shows two biasing arrangements for an N-channel JFET. At A, the gate is grounded through resistor R_2. The source resistor R_1 limits the current

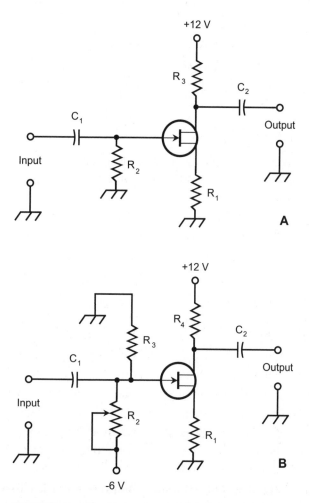

FIGURE 6-9 · Two methods of biasing an N-channel JFET. At A, fixed gate bias; at B, variable gate bias.

through the device. Resistors R_1 and R_2 determine the gate bias. The drain current I_D flows through R_3, producing a voltage across it. The AC output signal passes through C_2.

At B, the gate is connected through potentiometer R_2 to a negative DC voltage source. When we adjust this potentiometer, we vary the negative gate voltage E_G at the point between resistors R_2 and R_3. Resistor R_1 limits the current through the JFET. The drain current I_D flows through R_4, producing a voltage across it. The AC output signal passes through C_2.

In both of the circuits shown in Fig. 6-9, we connect the drain to a positive DC voltage source relative to ground. In the case of a P-channel JFET circuit, the polarities must be reversed.

The biasing arrangement in Fig. 6-9A is commonly used for weak-signal amplifiers, low-level amplifiers, and oscillators. The scheme shown in Fig. 6-9B is employed in certain power amplifiers having a substantial input signal. Typical JFET power-supply voltages are comparable to those with bipolar transistors. The voltage E_D between the drain and ground can range from approximately 3 to 50 V; most often it's 6 to 12 V.

How the JFET Amplifies

Figure 6-10 shows the relative drain (channel) current I_D as a function of the gate bias voltage E_G for a hypothetical N-channel JFET, assuming that the drain voltage E_D remains constant.

When E_G is fairly large and negative, the JFET operates in a state of pinchoff, so no current flows through the channel. As E_G gets less negative, the channel opens up and current begins flowing. As E_G gets still less negative, the channel grows wider and the drain current I_D increases. As E_G approaches the point where the source-gate (S-G) junction reaches forward breakover, the channel conducts as well as it possibly can; it's "wide open." If E_G gets more positive still, exceeding the forward breakover voltage and causing the S-G junction to conduct, some of the current in the channel leaks out through the gate. In most JFET circuits, we *do not* want that sort of leakage to occur.

The greatest amplification for weak signals occurs when we set E_G so that the curve in Fig. 6-10 has its steepest slope, as shown by the range marked X. In a high-power RF transmitting amplifier where the input signal is fairly strong, we'll often get the best results when we bias a JFET at or beyond pinchoff, in the range marked Y.

In a practical JFET amplifier circuit, the drain current passes through the drain resistor, as shown in Fig. 6-9A or Fig. 6-9B. Small fluctuations in E_G cause

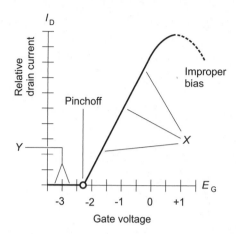

FIGURE 6-10 · Relative drain current as a function of gate voltage for a hypothetical N-channel JFET.

large changes in I_D, and these variations in turn produce wide swings in the DC voltage across R_3 (at A) or R_4 (at B). The AC part of this voltage goes through capacitor C_2, and appears at the output as a signal of much greater peak-to-peak voltage than that of the input signal at the gate. Therefore, the JFET operates as a *voltage amplifier*.

Drain Current versus Drain Voltage

In any JFET, the drain current I_D can be plotted as a function of drain voltage E_D for various values of gate voltage E_G. The resulting graph is called a *family of characteristic curves* for the device. Fig. 6-11 shows a family of characteristic curves for a hypothetical N-channel JFET.

Transconductance

As we've learned, the beta specification tells us how well a bipolar transistor can amplify an AC signal. When working with JFETs, engineers talk about *dynamic mutual conductance* or *transconductance* instead of beta.

Refer again to Fig. 6-10. Suppose that we set the gate voltage E_G at a certain value, resulting in a certain drain current I_D. If the gate voltage changes by a small amount ΔE_G, then the drain current will change by an increment ΔI_D. The transconductance g_{FS} equals the ratio of the change in the drain current to the change in the gate voltage. Mathematically, we have

$$g_{FS} = \Delta I_D / \Delta E_G$$

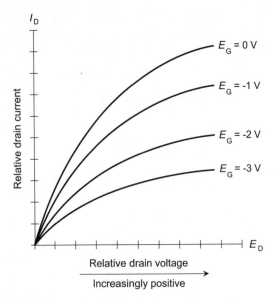

FIGURE 6-11 · A family of characteristic curves for a hypothetical N-channel JFET.

In graphical terms, the transconductance at any particular bias point translates to the slope of a line tangent to the curve of Fig. 6-10 at that point. We express transconductance in siemens, the same unit that we use for DC conductance.

As we can see from the graph of Fig. 6-10, the slope of a line tangent to the curve, and therefore the value of g_{FS}, varies as we move along the curve. When we bias the JFET beyond pinchoff in the region marked Y, the slope of a line tangent to the curve equals zero because it's a horizontal line. In the range marked Y, we'll observe no fluctuation in I_D when E_G changes by small amounts. We'll see a change in I_D with a change in E_G only when the channel conducts during at least some of an applied input signal's cycle.

The region of greatest transconductance corresponds to the portion of the curve marked X, where the curve has the greatest slope (ratio of "rise over run"). This region represents conditions where we can derive the most gain from the device. Because this part of the curve constitutes an almost straight line, we can expect to get excellent *linearity* (translating to minimal distortion) from any amplifier that we construct using the JFET, provided that we keep the input signal from getting so strong that it drives the device outside of range X during any part of the cycle.

If we bias the JFET beyond the range marked X, the slope of the curve decreases, and we can't get as much amplification as we do in the range marked X.

In addition, we can't expect the JFET to remain linear, because the curve does not constitute a straight line in the "improper bias" range. If we keep on biasing the gate of the N-channel JFET to greater and greater positive voltages, we reach the broken portion of the curve, arriving at the zone where the S-G junction goes past forward breakover and begins to draw current from the channel.

PROBLEM 6-3

Refer to Fig. 6-11. Imagine that we operate the device described by this family of curves with a DC gate voltage $E_G = -1$ V. Suppose that we define and measure the "effective resistance" R of this device by dividing the DC drain voltage by the DC drain current. What happens to R as the drain voltage E_D increases?

SOLUTION

Mathematically, we can calculate the value of R using the equation

$$R = E_D / I_D$$

This ratio is smallest at small drain voltages, and gradually increases as E_D increases, a result of the fact that the drain current (or current through the device) increases less and less rapidly as the drain voltage increases.

PROBLEM 6-4

Look again at Fig. 6-10 on page 189. What happens to the transconductance as the gate voltage starts at pinchoff and then grows less negative, passing through 0 and then increasing positively?

SOLUTION

The transconductance that we observe for any specific gate voltage E_G equals the slope of the curve at the point corresponding to that value of E_G. At first, the slope appears large and positive, because the curve ramps upward (the region marked X). When the gate voltage reaches a certain point, the transconductance starts to decrease, and reaches 0 when E_G is a little less than +1 V. Beyond this point, the transconductance becomes negative because a further positive increase in E_G causes the drain current I_D to decline.

TIP *From your algebra or precalculus courses, recall that in a rectangular coordinate graph, such as Fig. 6-10, curves that "ramp up locally" as you move toward the right have positive local slope; curves that "ramp down locally" as you move toward the right have negative local slope.*

Metal-Oxide FETs

The acronym MOSFET (pronounced "MOSS-fet") stands for *metal-oxide-semiconductor field-effect transistor*. This type of component can be constructed with a channel of N type material, or with a channel of P type material. We call the former type an *N-channel MOSFET* and the latter type a *P-channel MOSFET*. Figure 6-12A is a simplified cross-sectional drawing of an N-channel MOSFET. Figure 6-12B shows the schematic symbol. The P-channel cross-sectional drawing and symbol appear at C and D in Fig. 6-12.

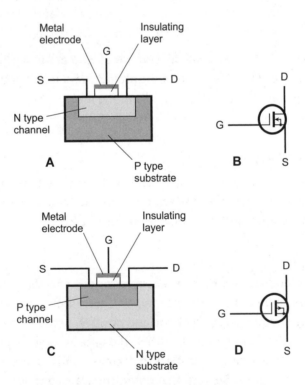

FIGURE 6-12 · Pictorial diagram of an N-channel MOSFET (A), the schematic symbol for an N-channel MOSFET (B), pictorial diagram of a P-channel MOSFET (C), and the schematic symbol for a P-channel MOSFET (D).

The Insulated Gate

When semiconductor engineers first conceived and developed the MOSFET, they called it an *insulated-gate FET* or IGFET. This expression perhaps describes the device better than the currently accepted term does! The gate electrode is actually insulated, by a thin layer of dielectric material, from the channel.

The input impedance for a MOSFET exceeds that of a JFET when we apply an input signal at the gate electrode. In fact, the gate-to-source (G-S) resistance of a typical MOSFET compares quite favorably to the *leakage resistance* of a well-designed capacitor. It's so large that we can usually consider it "practically infinite."

Figure 6-13 shows a family of characteristic curves for a hypothetical N-channel MOSFET. Note that the curves rise steeply at first for relatively small values of drain voltage E_D; but as E_D increases beyond a certain threshold, the curves level off more quickly than they do for a JFET.

TIP *One of the most serious shortcomings of MOSFETs is the fact that they're easily destroyed by electrostatic discharge. When circuits containing MOS devices are built or serviced, technicians must use special equipment to ensure that their hands don't acquire electrostatic charges. If any stray discharge occurs through*

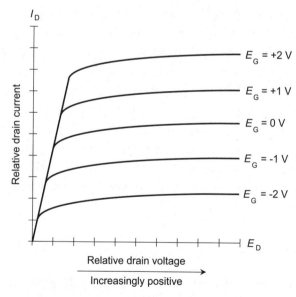

FIGURE 6-13 · A family of characteristic curves for a hypothetical N-channel MOSFET.

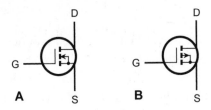

FIGURE 6-14 · At A, the schematic symbol for an N-channel enhancement-mode MOSFET. At B, the schematic symbol for a P-channel enhancement-mode MOSFET.

the thin, fragile dielectric layer inside a MOSFET, the resulting current can, and often does, permanently ruin the device.

Depletion Mode versus Enhancement Mode

In a JFET, the channel conducts with zero gate bias, that is, when the gate has the same voltage as the source ($E_G = 0$). As E_G increases (negatively for an N-channel device and positively for a P-channel device), the depletion region grows and the charge carriers must pass through a narrowed channel. We call this condition the *depletion mode*. Some MOSFETs can also function in the depletion mode. The drawings and schematic symbols of Fig. 6-12 depict the internal construction and schematic symbols for a *depletion-mode MOSFET*.

Metal-oxide-semiconductor technology allows for an alternative electrical environment that differs substantially from the depletion mode. An *enhancement-mode MOSFET* has a pinched-off channel at zero bias. We must apply a gate bias voltage, E_G, to create a channel in this type of device. If $E_G = 0$, then $I_D = 0$ in the absence of signal input. We apply gate bias and signals to enhance (widen), rather than to deplete (narrow), the channel through the semiconductor material.

Figure 6-14 shows the schematic symbols for N-channel and P-channel enhancement-mode MOSFETs. The right-hand vertical lines inside the symbols are broken, rather than solid as in the symbol for the depletion-mode MOSFET. This difference allows us to distinguish between the two types of devices when we see them in circuit diagrams.

Basic FET Circuits

Three general circuit configurations exist for FETs. They're the equivalents of the common-emitter, common-base, and common-collector bipolar-transistor circuits, and they break down as follows:

1. The *common-source circuit*, in which we ground the source for AC signals

2. The *common-gate circuit*, in which we ground the gate for AC signals

3. The *common-drain circuit*, in which we ground the drain for AC signals

Common Source

In a *common-source circuit*, we apply the input signal to the gate, as shown in Fig. 6-15. This diagram shows an N-channel JFET, but we could substitute an N-channel, depletion-mode MOSFET and get the same results in a practical circuit. We could also use an N-channel, enhancement-mode MOSFET and add an extra resistor between the gate and the positive power supply terminal.

For P-channel devices, the power supply would have to provide a negative voltage rather than a positive voltage. Otherwise the circuit details would correspond to those shown in Fig. 6-15.

Capacitor C_1 and resistor R_1 place the source at signal ground while elevating the source above ground for DC. The AC signal enters through capacitor C_2. Resistors R_1 and R_2 provide bias for the gate. The AC signal passes out of the circuit through capacitor C_3. Resistor R_3 keeps the output signal from shorting through the power supply.

The circuit of Fig. 6-15 provides a good starting point for the design of weak-signal amplifiers and low-power oscillators, especially in RF systems. The common-source arrangement provides the greatest gain of the three FET circuit configurations. The output wave is inverted with respect to the input wave.

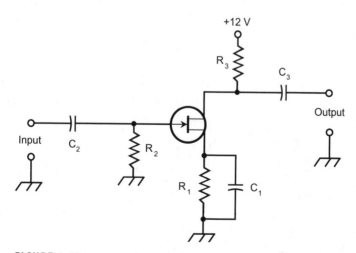

FIGURE 6-15 · Common-source N-channel configuration. Illustration for Problem 6-5.

Common Gate

The *common-gate circuit* (Fig. 6-16) has the gate at signal ground. We apply the input signal to the source. This illustration shows an N-channel JFET. For other types of FETs, the same considerations apply as in the case of the common-source circuit. An N-channel enhancement-mode device requires a resistor between the gate and the positive supply terminal. For P-channel devices, we reverse the polarity of the power supply.

The DC bias for the common-gate circuit resembles that for the common-source arrangement, but the signal follows a different path. The AC input signal enters through capacitor C_1. Resistor R_1 keeps the input from shorting to ground. Resistors R_1 and R_2 provide the gate bias. Capacitor C_2 places the gate at signal ground while allowing DC bias voltage to exist on that electrode. The output signal exits the circuit through C_3. Resistor R_3 keeps the output signal from shorting through the power supply.

The common-gate arrangement produces less gain than its common-source counterpart. However, a common-gate amplifier is far less likely than a common-source amplifier to break into unwanted oscillation. The output wave occurs in phase coincidence with the input wave.

Common Drain

Figure 6-17 shows a *common-drain circuit*, in which we place the drain at signal ground. Engineers sometimes call this circuit a *source follower* because

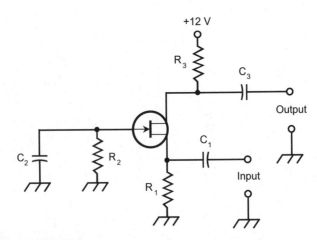

FIGURE 6-16 · Common-gate N-channel configuration. Illustration for Problem 6-6.

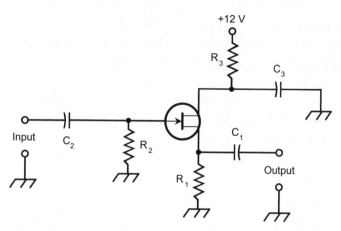

FIGURE 6-17 · Common-drain N-channel configuration.

the output signal waveform "follows" the instantaneous voltage at the source electrode.

To use an FET in the common-drain configuration, we bias it for DC in the same way as we do in the common-source and common-gate circuits. The circuit shown in Fig. 6-17 employs an N-channel JFET, but we can substitute any other kind of FET, reversing the polarity for P-channel devices. Enhancement-mode MOSFETs require a resistor between the gate and the positive supply terminal (or the negative terminal for a P-channel device).

The input signal passes through capacitor C_2 to the gate. Resistors R_1 and R_2 provide gate bias. Resistor R_3 limits the current. Capacitor C_3 keeps the drain at signal ground. Fluctuating DC (the channel current) flows through R_1 as a result of the input signal; this current causes a fluctuating DC voltage to appear across R_1. We take the AC output through C_1.

The output wave of the common-drain circuit exists in phase coincidence with the input wave. This circuit, like the common-collector arrangement, can serve as a low-cost alternative to a transformer, especially in RF applications.

Still Struggling

In the preceding discussions (accompanied by Figs. 6-15, 6-16, and 6-17), you might get the idea that we can directly interchange JFETs and MOSFETs in practical circuits. Sometimes we can do that, but not always. Although the circuit diagrams for JFET and depletion-mode MOSFET devices look identical when we don't specify actual component values, there's usually a difference in the optimum resistances and capacitances. These optimum values can vary not only between JFETs and MOSFETs, but also between depletion-mode MOSFETs and enhancement-mode MOSFETs. Specific design examples would take us beyond the scope of this course, but you should remember that JFETs and MOSFETs are sometimes, but not always, directly interchangeable.

PROBLEM 6-5

In the common-source circuit (Fig. 6-15 on page 195), what would happen to the output signal if R_2 were to fail in such a way as to cause its resistance to drop to zero? What would happen to the gate bias?

SOLUTION

Resistor R_2 exists *in shunt* with (that is, in parallel with, or across) the input signal path. This resistor, along with R_1, provides bias for the gate with respect to DC ground. If R_2 were to become a short circuit, the input signal would disappear from the gate because it would short directly to ground. With no input signal, we would get no output signal. In addition, this situation would alter the DC gate bias.

PROBLEM 6-6

In the common-gate circuit (Fig. 6-16 on page 196), what would happen to the drain current if resistor R_1 were to open? What would happen to the output signal?

SOLUTION

Resistor R_1 exists in series with the current path through the channel. Therefore, if R_1 were to open up, the DC through the channel (and therefore at the drain) would cease to flow. This interruption would prevent the FET from amplifying the signal. A little of the input signal might leak through to the output, but even if it did, the output signal could be no stronger than the input signal.

Integrated Circuits

Most *integrated circuits* (ICs) look like plastic boxes with protruding metal pins. The schematic symbol for an IC is a triangle or rectangle with the component designator written inside. We'll encounter two types of ICs in common practice.

1. A *linear IC* processes analog signals. The term "linear" arises from the fact that the instantaneous output constitutes a straight-line mathematical function of the instantaneous input.

2. A *digital IC* comprises circuits called *logic gates* that perform *binary* (two-state) operations at high speeds. Logic gates are not linear; they always function in either the high state (logic 1) or the low state (logic 0).

Compactness

Integrated-circuit devices and systems are far more compact than equivalent circuits made from *discrete components* (individual resistors, capacitors, diodes, and transistors). More complex circuits can be built, and kept down to a reasonable size, using ICs as compared with discrete components. That's why, for example, we'll find notebook computers in department stores today that offer features more advanced than the most powerful "supercomputers" built in the middle of the last century.

High Speed

In an IC, the interconnections among components are physically tiny, making high switching speeds possible. Electric currents travel fast, but not instantaneously. The faster the charge carriers can get from one component to another, the more operations can occur per unit of time, and the less time it takes for the device to perform a fixed number of operations.

Low Power Requirement

Integrated circuits usually require less power than equivalent discrete-component circuits. This consideration attains critical importance in battery-powered electronic systems. Integrated circuits draw less current than equivalent circuits built from discrete resistors, capacitors, diodes, and transistors. Therefore, ICs produce less heat and offer better efficiency, minimizing problems that plague equipment that gets hot with use. Such problems can include *frequency drift*, the generation of excessive *internal electrical noise*, reduced reliability, and total system failure (also called *cat-astrophic failure*).

TIP *Despite their modest power needs, certain types of ICs generate enough heat, in proportion to their small physical size, to require cooling fans, heatsinks, or other schemes to keep them from getting too hot. The microprocessor in a high-end personal computer provides a good example.*

Reliability

Systems using ICs fail less often, per component-hour of use, than systems comprising discrete components, mainly because all interconnections are sealed within an IC case, preventing corrosion or the intrusion of dust and water vapor. The reduced failure rate translates into less downtime.

Ease of Maintenance

Integrated-circuit technology keeps service costs to a minimum because repair procedures are simple when failures occur. Many systems use sockets for ICs, and replacement involves nothing more than identifying the faulty IC, unplugging it, and plugging in a new one. Technicians use special desoldering equipment to service circuit boards that have ICs soldered directly to the foil.

Modular Construction

Modern IC appliances employ *modular construction*, in which specific circuits, devices, or systems exist on dedicated circuit boards called *cards*. Individual ICs perform defined functions within a card. The card, in turn, fits into a socket on a larger circuit board. Technicians use computers, programmed with custom-ized software, to locate a faulty card in a sophisticated system. The card can be pulled and replaced, getting the system back to the end user in the shortest possible time. That process helps to keep customers happy (or at least mini-mizes customer frustration with the service department).

Inductors Impractical

Devices using ICs must generally function without inductors because induc-tances cannot easily be fabricated onto silicon chips. Resistance-capacitance (*RC*) circuits can perform most operations that inductance-capacitance (*LC*) circuits can do. Therefore, most tasks that call for inductances can be re-designed to use resistances instead, which lend themselves easily to semicon-ductor fabrication. Alternatively, engineers can sometimes get an entire IC to exhibit inductive reactance at specific frequencies, and thereby replace a conventional coil in a larger circuit.

Mega-Power Impossible

High-power amplifiers cannot, in general, be built onto semiconductor chips. High power necessitates a certain minimum physical bulk and mass to allow the conduction and radiation of excess heat energy. Power transistors and, in some systems, vacuum tubes are generally employed for high-power amplification.

Operational Amplifiers

An *operational amplifier* (often called an *op amp* for short) is a linear IC that produces gain over a wide range of signal input frequencies. An op amp has two inputs and one output.

When we apply a signal to the *noninverting input* of an op amp, the output wave occurs in phase coincidence with the input wave. When we apply a signal to the *inverting input*, the output wave comes out in phase opposition with respect to the input wave. (In the case of a pure sine wave, this inversion constitutes the equivalent of a 180° shift in phase.)

An op amp has two power supply connections, one for the emitters of the microscopic internal transistors (V_{ee}) and one for the collectors (V_{cc}). Engineers usually symbolize op amps as triangles. The inputs, output and power supply connections are drawn as lines emerging from the triangle.

External resistors determine the gain of an op amp. Normally, we connect a resistor between the output and the inverting input, giving us the so-called *closed-loop configuration* with *negative feedback* that occurs out of phase with the input signal, causing the op amp's gain to decrease relative to the no-feedback situation (called the *open-loop configuration*). Figure 6-18 illustrates a basic closed-loop amplifier using an op amp.

If we connect a resistor between the output and the noninverting input rather than between the output and the inverting input, positive feedback takes place. If we choose certain values for the feedback resistor in this scenario, we can get the op amp to oscillate.

When we place a resistance-capacitance (RC) combination in the negative feedback loop of an op amp, the amplification factor varies with the frequency. We can thereby get a *lowpass response* (gain decreases as the frequency increases), a *highpass response* (gain increases as the frequency increases), a *resonant peak* (gain reaches a maximum at a specific frequency), or a *resonant notch* (gain reaches a minimum at a specific frequency) using an op amp and various RC

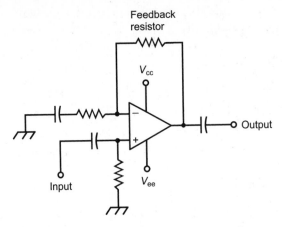

FIGURE 6-18 · A closed-loop op amp circuit employs negative feedback to control the gain.

feedback arrangements. Figure 6-19 illustrates these responses as gain-versus-frequency graphs.

TIP   *If we employ an RC combination in a positive-feedback op-amp loop with the intent of building an oscillator, the resistance and capacitance values determine the frequency at which the circuit oscillates.*

Voltage Regulator

A *voltage regulator* is a linear IC that controls, or governs, the output voltage of a power supply. Most precision electronic equipment requires operating voltages that remain within a narrow range. We can find voltage-regulator ICs with almost any voltage and current rating we'll ever need in a practical electronic system. Typical voltage regulator ICs have three terminals. Casual observers sometimes mistake voltage-regulator ICs for large transistors.

Timer

A *timer* IC is form of oscillator that produces a delayed output, with the delay being variable to suit the needs of a particular device. The delay is generated by counting the number of oscillator pulses. The length of the delay can be adjusted using external potentiometers and capacitors.

Multiplexer and Demultiplexer

A *multiplexer* is a linear IC that combines several different signals in a single communications channel. A *demultiplexer* separates the individual signals from

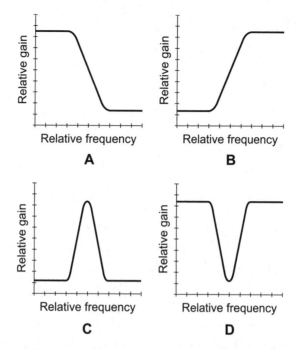

FIGURE 6-19 · Gain-versus-frequency response curves. At A, lowpass; at B, highpass; at C, resonant peak; at D, resonant notch.

a channel containing multiple signals. A *multiplexer/demultiplexer* can perform either of these functions, or both at the same time.

Comparator

A *comparator* IC has two inputs. The device literally compares the voltages at two inputs called A and B. If the input voltage at A significantly exceeds the input voltage at B, the comparator produces a DC output of about +5 V, representing logic 1, or the high binary state. If the input voltage at A is less than or equal to the input voltage at B, the IC produces about +2 V DC at the output, representing logic 0, or the low binary state.

Comparators are employed to actuate, or *trigger*, other devices, such as relays and electronic switching circuits. Some comparators can switch between low and high states at high speed; others work slowly. Some have low input impedance; others have high impedance. Some are intended for audio or low-frequency use; others are fabricated for video or high-frequency applications.

Random-Access Memory (RAM)

A *random-access memory* (RAM) chip is a digital IC that stores binary data in arrays. The data can be addressed from anywhere in the matrix. Data is easily changed and stored back in RAM. With most RAM chips, data vanishes immediately on power-down unless some provision is made for *memory backup*. Any memory medium whose content disappears when we remove the power constitutes so-called *volatile memory*. If the data remains intact even when we remove power for an extended period, we have *nonvolatile memory*.

Read-Only Memory (ROM)

Read-only memory (ROM) can be accessed in whole or in part, but not written over in the course of normal operation. A standard ROM chip, which constitutes a digital device, is programmed at the factory. This permanent programming is known as *firmware*. There are also ROMs that you can program and reprogram yourself.

An *erasable programmable ROM* (EPROM) is a digital ROM chip that engineers and technicians can reprogram by exposing the semiconductor surface to ultraviolet (UV) radiation. The IC must be physically removed from its circuit card, exposed to the UV for several minutes, reprogrammed by following a special procedure, and then put back in the card.

There are EPROMs that engineers and technicians can erase by electrical means, without the need for UV-generating devices. Such an IC is called an EEPROM, which stands for *electrically erasable programmable ROM*. This type of IC can remain on a circuit board during the reprogramming process.

TIP *Some EPROM chips can suffer data erasure or corruption if exposed to X rays. This fact might explain why some people claim to have experienced malfunctions in notebook computers and other digital devices after they've passed through airport security scanners.*

PROBLEM 6-7

What would happen if the feedback resistor in the circuit of Fig. 6-18 on page 202 were to open up?

SOLUTION

In that case, the negative feedback would drop to zero, and the op amp would operate at its maximum possible gain.

PROBLEM 6-8

What would happen if we were to replace the feedback resistor in the circuit of Fig. 6-18 with a direct connection?

SOLUTION

If we connect the inverting output directly to the input, we'll maximize the amount of negative feedback, causing a dramatic decrease in the gain of the device and in the output signal amplitude. In the extreme, this action might cause the output signal to vanish altogether.

QUIZ

Refer to the text in this chapter if necessary. A good score is eight correct. The answers are listed in the back of the book.

1. If we expect to observe any collector current in a common-emitter PNP bipolar transistor circuit with the base biased at +1.2 V DC relative to the emitter and the collector biased at −18 V DC relative to ground, an AC input signal at the emitter-base junction must reach or exceed a certain

    A. negative peak voltage for *part* of the cycle.
    B. positive peak voltage for *part* of the cycle.
    C. negative peak voltage for *all* of the cycle.
    D. positive peak voltage for *all* of the cycle.

2. In the source-follower circuit of Fig. 6-20, what function does the component marked X perform?

    A. It helps to keep the circuit oscillating.
    B. It keeps the drain at signal ground while allowing a DC voltage to exist there.
    C. It allows the signal to exist at the drain while preventing any DC voltage from existing there.
    D. It keeps the signal from shorting through the power supply.

3. What's the static forward current transfer ratio H_{FE} in a PNP bipolar transistor where a base current of 5.0 mA produces a collector current of 20 mA?

    A. 100
    B. 2.0
    C. 0.25
    D. 4.0

4. Which of the following amplifier circuit configurations can produce the most gain?

    A. Common-gate
    B. Common-source
    C. Common-collector
    D. Common-drain

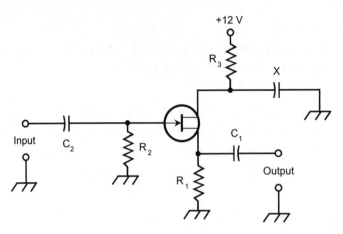

FIGURE 6-20 · Illustration for Quiz Question 2.

5. Figure 6-21 portrays the function of relative drain current (I_D) versus gate voltage (E_G) for a hypothetical N-channel JFET. Which of the following DC gate bias voltages will give us the highest transconductance?

  A. $E_G = -3.00$ V
  B. $E_G = -2.25$ V
  C. $E_G = -1.00$ V
  D. $E_G = +0.75$ V

6. In Fig. 6-21, the downward-bending part of the curve at the extreme upper right end of the graph represents a condition of

  A. pinchoff.
  B. current at the S-G junction.
  C. reverse bias.
  D. oscillation.

7. Under no-signal conditions in an NPN bipolar transistor at zero base bias and the collector at +6 V DC, we should expect

  A. a small collector current to flow in the forward direction.
  B. a large collector current to flow in the reverse direction.
  C. no collector current to flow.
  D. a small, alternating collector current to flow.

8. In some applications, we can use an emitter-follower circuit as

  A. a voltage regulator.
  B. an oscillator.
  C. an alternative to a transformer.
  D. a weak-signal amplifier.

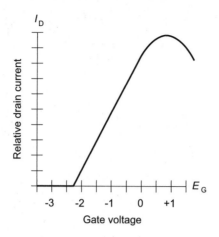

FIGURE 6-21 • Illustration for Quiz
Questions 5 and 6.

9. **If the alpha of a certain transistor equals 0.800 at 12.5 MHz, what's the beta of that same transistor at 12.5 MHz?**

   A. We need more information to answer this question.

   B. 0.67

   C. 1.25

   D. 4.00

10. **If the alpha of a certain transistor equals 0.800 at 12.5 MHz, what's the alpha cutoff of that same transistor at 12.5 MHz?**

   A. We need more information to answer this question.

   B. 8.84 MHz

   C. 10.0 MHz

   D. 17.7 MHz

chapter *7*

Signal Amplifiers

A *signal amplifier* increases the amplitude (the current, voltage, or power level) of an audio-frequency (AF) or radio-frequency (RF) signal. In this chapter, we'll learn how the most common types of signal amplifiers work.

CHAPTER OBJECTIVES

In this chapter you will

- Express voltage, current, and power levels in decibel form.
- Compare bipolar-transistor and FET amplifier circuits.
- See how biasing affects the behavior of an amplifier.
- Learn the characteristics of the basic amplifier classes.
- Compare RF and AF amplifiers.
- Compare broadband and tuned RF amplifiers.

Amplification Factor

We can quantify the extent to which a circuit increases signal strength in terms of an *amplification factor*, expressed in *decibels* (dB). The amplification factor can differ as defined for voltage, current, and power.

The Decibel

Humans perceive most variable quantities in a nonlinear manner. Scientists and engineers have devised the *decibel system*, in which amplitude variations are expressed according to the *logarithm* of the actual signal strength. *Gain* produces positive decibel values; *loss* yields negative decibel values.

> **TIP**   *An amplitude change of ±1 dB roughly equals the smallest extent to which we can detect an* anticipated, sudden *change in signal strength. If we* do not *anticipate or expect any change in the strength of a signal, then the smallest noticeable difference, if the transition occurs suddenly, is about ±3 dB.*

Voltage Decibels

Imagine an amplifier with an RMS AC input voltage of E_{in} and an RMS AC output voltage of E_{out}, both in the same units. Also, assume that the input impedance is identical to the output impedance. In this case, we can calculate the *voltage gain* of the circuit, in decibels, using the formula

$$\text{Gain (dB)} = 20 \log_{10} (E_{out}/E_{in})$$

Current Decibels

We can calculate the *current gain* of a circuit exactly as we do with voltage gain. If I_{in} represents the RMS AC input current and I_{out} represents the RMS AC output current specified in the same units, and if the input impedance equals the output impedance, then

$$\text{Gain (dB)} = 20 \log_{10} (I_{out}/I_{in})$$

Any *passive* device or circuit (one that has no external power source such as a battery) that produces voltage gain must inevitably yield a current loss, and vice-versa. An example is an AC transformer. *Active* devices or circuits (ones that have external power sources and employ components such as transistors or op amps) can, in some cases, produce gain for both voltage and current.

TIP *When a circuit exhibits voltage gain and current gain, the decibel figures will differ if the output impedance is higher or lower than the input impedance, altering the ratio of voltage to current. The two figures will equal each other, if and only if, the input impedance is identical to the output impedance.*

Power Decibels

We can calculate the *power gain* of a device, circuit, or system in decibels using the formula

$$\text{Gain (dB)} = 10 \log_{10} (P_{out}/P_{in})$$

where P_{out} represents the output signal power and P_{in} represents the input signal power, specified in the same units. This formula works regardless of the input and output impedances. They can differ substantially and we'll still get valid results.

PROBLEM 7-1

Suppose that the RMS AC input voltage to a circuit equals 10 millivolts (mV) and the RMS AC output voltage equals 70 mV. What's the voltage gain? Assume that the input and output impedances are identical.

SOLUTION

We can assign E_{in} = 10 and E_{out} = 70, and then use the previously defined formula for voltage gain to get

$$\text{Gain (dB)} = 20 \log_{10} (E_{out}/E_{in})$$
$$= 20 \log_{10} (70/10)$$
$$= 20 \log_{10} 7.0$$
$$= 20 \times 0.845$$
$$= +16.9 \text{ dB}$$

We should round this answer off to +17 dB, because our input data is accurate to only two significant figures.

PROBLEM 7-2

If the RMS AC input current to a circuit is 250 microamperes (µA) and the output is 100 milliamperes (mA), what's the current gain? Assume that the input and output impedances are identical.

SOLUTION

First, we must convert the input and output current to the same units. Let's use milliamperes. Then we have $I_{in} = 0.250$ and $I_{out} = 100$, and we can use the previously defined formula for current gain to get

$$\text{Gain (dB)} = 20 \log_{10} (I_{out} / I_{in})$$
$$= 20 \log_{10} (100 / 0.250)$$
$$= 20 \log_{10} 400$$
$$= 20 \times 2.602$$
$$= +52.04 \text{ dB}$$

We should round this result off to 52.0 dB, because the accuracy of our input data extends to only three significant figures.

PROBLEM 7-3

Suppose that we apply a certain amount of signal power to an *RF attenuator* (a circuit especially designed to reduce the strength of a signal by dissipating some of it in a network of resistors). We get *exactly* 1/15 as much power at the output as we provide at the input. What's the power gain in decibels? What's the power loss in decibels?

SOLUTION

We have a ratio here, not actual power values. Nevertheless, we can calculate the power gain and loss factors; we only need to know the ratio. Let's assign an input power value $P_{in} = 15$ units and $P_{out} = 1$ unit. (We could use any other numbers, as long as they exist in the ratio of 15 to 1.) Then, using the previously defined formula for power gain, we get

$$\text{Gain (dB)} = 10 \log_{10} (P_{out} / P_{in})$$
$$= 10 \log_{10} (1/15)$$
$$= 10 \log_{10} 0.06667$$
$$= 10 \times (-1.176)$$
$$= -11.76 \text{ dB}$$

Because the attenuation factor is exact, we can leave this figure alone. (An experienced engineer might round it off to −12 dB, allowing for a bit of justifiable doubt!) That's the power gain of the attenuator. We can also say that the circuit produces 12 dB of power loss.

Basic Amplifier Circuits

In general, amplifiers must use active components, such as bipolar transistors, FETs, or op-amp ICs. A simple AC transformer, which constitutes a passive device, can increase the deliverable current or voltage if the input and output impedances differ. But it can't produce an output signal that has more power than the input signal under any circumstances.

Amplifier Using a Bipolar Transistor

Figure 7-1 is a schematic diagram of a generic NPN bipolar-transistor amplifier. The input signal passes through a capacitor to the base. Resistors provide base bias. In this amplifier, the capacitors must have values large enough to allow the AC signal to pass with ease, but they should not be much larger than the minimum necessary for this purpose. The ideal capacitance values depend on the design frequency of the amplifier, and also on the impedances at the input and output. In general, as the frequency and/or circuit impedance increase, we need less capacitance. The input and output resistor values should be comparable with the resistive parts of the input and output impedances, respectively.

Amplifier Using a JFET

Figure 7-2 shows a generic amplifier that employs an N-channel JFET. All FETs exhibit high input impedance, so we can use a relatively small-value input

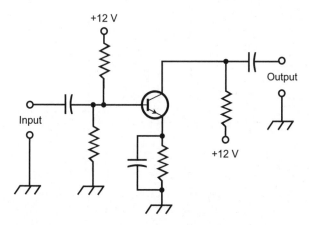

FIGURE 7-1 · Generic diagram of an NPN bipolar-transistor amplifier circuit.

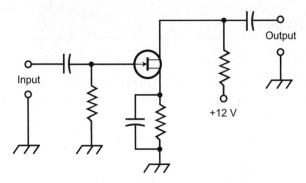

FIGURE 7-2 · Generic diagram of an N-channel FET amplifier circuit.

capacitor. If we use a MOSFET instead of a JFET, the input impedance will be larger so the input capacitance can be tiny indeed—sometimes less than 1 pico-farad (pF) for weak-signal RF amplifiers. Resistor values depend on the input and output impedances.

Amplifier Using an Op Amp

Figure 7-3 illustrates a simple audio amplifier using an op-amp IC. In this case, we have an *inverting amplifier*, meaning that the output wave exists in phase opposition with respect to the input wave. The *RC* combinations in the input

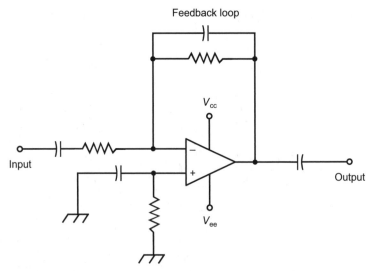

FIGURE 7-3 · In an op-amp circuit, we can use resistors and capacitors to "tailor" the gain and frequency response.

and the feedback loop serve to "tailor" the gain and the frequency response. In general, larger capacitance values favor low frequencies, and small capacitances favor higher frequencies.

Amplifier Classes

Engineers categorize signal amplifier circuits as *class A*, *class AB*, *class B*, or *class C*, depending on the way the active devices are biased. Each class has unique characteristics and applications.

Class A

Weak-signal amplifiers, such as the kind used in an audio microphone pream-plifier or in the first stage of a sensitive radio receiver, always operate in the *class-A* mode. This type of amplifier exhibits excellent *signal linearity*, meaning that the output-signal wave shape (or waveform) constitutes a faithful, but magnified, reproduction of the input-signal waveform. If we input a sine-wave signal to an amplifier with excellent signal linearity, then we can expect to get a sine-wave signal out.

In order to obtain class-A operation with a bipolar transistor, the bias must be such that, with no signal input, the device operates near the middle of the straight-line portion of the collector current (I_C) versus base current (I_B) curve. Figure 7-4 illustrates this state of affairs for a bipolar transistor. With a JFET or

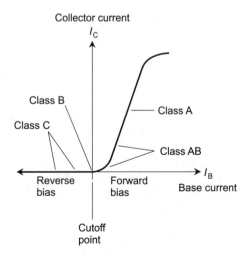

FIGURE 7-4 • Classes of amplification for a bipolar transistor.

MOSFET, the bias must be such that, with no signal input, the device operates near the middle of the straight-line part of the drain current (I_D) versus gate voltage (E_G) curve. Figure 7-5 shows this situation for a JFET.

Class AB

When we bias a bipolar transistor slightly above the point where the no-signal base current becomes zero (cutoff), or when we bias an FET slightly above the point where the no-signal gate current becomes zero (pinchoff), the input signal drives the device into the nonlinear part of the operating curve. Then we have *class-AB operation*. Figure 7-4 shows this situation for a bipolar transistor, and Fig. 7-5 shows it for a JFET.

In a class-AB amplifier, the input signal might cause the device to go into cutoff or pinchoff for a small part of the cycle. The actual bias point, along with the peak-to-peak strength of the input signal, determines whether or not "partial-cycle cutoff" or "partial-cycle pinchoff" takes place. If the bipolar transistor or FET never goes into cutoff or pinchoff during any part of the signal cycle, we have a *class-AB₁ amplifier*. If the device goes into cutoff or pinchoff for any part of the cycle, we have a *class-AB₂ amplifier*. Class-AB operation works quite well in RF power-amplifier (PA) circuits.

TIP   *In any class-AB amplifier, the output waveform differs in shape from the input waveform. However, if we modulate the signal, such as in a voice radio transmitter,*

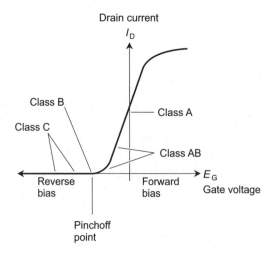

FIGURE 7-5 · Classes of amplification for a JFET.

the data impressed on the signal will emerge undistorted anyway. In other words, the modulation envelope *(the function of instantaneous signal strength versus time) will remain undistorted, even though the signal waveform itself changes considerably from input to output.*

Class B

When we bias a bipolar transistor exactly at cutoff, or when we bias an FET exactly at pinchoff, we obtain an amplifier that operates in the class-B mode. These operating points are labeled on the curves in Figs. 7-4 and 7-5. In a *class-B amplifier*, we observe no collector or drain current in the absence of a signal at the input. This characteristic saves energy compared with class-A and class-AB circuits because a class-B amplifier doesn't have to "do any work" when it doesn't get an input signal. When an input signal appears, current flows in the device during half of the cycle.

The class-B scheme, like the class-AB$_1$ or class-AB$_2$ modes, performs effectively for RF power amplification. The output wave has a shape that differs greatly from that of the input wave, so we observe harmonics in addition to the fundamental frequency at the amplifier output. We can overcome this problem by incorporating a resonant inductance-capacitance (LC) circuit in the output, tuned to the fundamental frequency. If we modulate the signal in a class-B amplifier, the modulation envelope comes out undistorted, even though the signal waveform changes.

Push-Pull

We can connect two bipolar transistors or FETs in a class-AB or class-B circuit, using one device to amplify the positive half of the cycle and the other device to amplify the negative half. In this way, we can eliminate the waveform distortion inherent in class-AB or class-B circuits. Engineers call this arrangement a *push-pull amplifier* (Fig. 7-6). A push-pull amplifier effectively "cancels" signal output energy at *even harmonics* of the input frequency (multiples of 2, 4, 6, and so on). Push-pull operation has little or no effect on output energy at *odd harmonics* (multiples of 3, 5, 7, and so on).

Class C

We can bias a bipolar transistor or FET beyond the cutoff or pinchoff point, and it can work as a PA if we provide enough input-signal power to overcome the

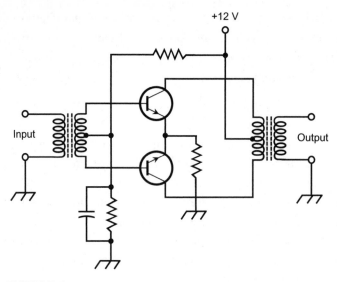

FIGURE 7-6 · Generic diagram of a push-pull amplifier.

bias and cause the device to conduct during part of the cycle. We call this type of system a *class-C amplifier*. Approximate operating points for class C are labeled in Figs. 7-4 and 7-5.

A class-C RF amplifier has poor linearity (in other words, it's *nonlinear*) for signal envelopes in which the amplitude varies, such as we find in a standard *amplitude-modulated* (AM) signal. The class-C circuit offers distortion-free operation only for a signal whose amplitude remains constant, or else, has only two levels (called on/off, high/low, or mark/space). Examples of such modes include "continuous-wave" (CW) *radiotelegraphy, frequency-shift keying* (FSK), *phase-shift keying* (PSK), and *frequency modulation* (FM).

TIP *A class-C RF PA needs substantial input power in order to overcome the cutoff or pinchoff bias at its base or gate. When properly operated, however, a class-C amplifier offers high efficiency. Some FM broadcast transmitters employ this mode.*

Efficiency and Drive

In a PA, we define the *efficiency* as the ratio of the useful *AC signal power output* to the total *DC power input*. High efficiency ensures minimal generation of heat in the transistors, and also prolongs their useful lives. We define the *drive* as the AC signal input strength, usually in terms of power.

DC Power Input

The DC power input (P_{in}), in watts, to a bipolar-transistor amplifier circuit equals the product of the collector current (I_C) in amperes and the collector voltage (E_C) in volts. For an FET, the DC power input equals the product of the drain current (I_D) and the drain voltage (E_D). Mathematically, we have

$$P_{in} = E_C I_C$$

for bipolar-transistor power amplifiers, and

$$P_{in} = E_D I_D$$

for FET power amplifiers.

In some circuits, we observe considerable DC power input even when we apply no signal at the input. In a class-A amplifier, for example, the average DC power input remains constant whether an input signal exists or not, assuming that the input signal doesn't get so strong that it causes the circuit to go into a state of nonlinearity. In a class-AB$_1$ or class-AB$_2$ amplifier, we observe a small DC power input with no signal input, and significant DC power input when we apply a signal. In class-B and class-C amplifiers, the DC power input equals zero with no input signal; we obtain significant DC power input only if we apply a signal of sufficient strength.

Signal Power Output

When we don't provide an amplifier with any input signal, we get nothing at the output, so the *signal power output* (P_{out}) equals zero. This situation holds true for all classes of amplification. In general, as we increase the signal input amplitude, the power output of a power amplifier also increases—up to a certain point.

TIP   *We can't directly measure the signal power output from an amplifier using a DC instrument. We must employ a specialized AC wattmeter designed to provide true power readings at the signal frequency.*

Definition of Efficiency

The *efficiency* (*eff*) of a PA equals the ratio of the useful signal power output to the DC power input. The formula is

$$eff = P_{out} / P_{in}$$

This ratio always lies somewhere between 0 and 1. Alternatively, we can express amplifier efficiency as a percentage between 0% and 100% ($eff_{\%}$) using the formula

$$eff_{\%} = (100P_{out}/P_{int})\%$$

Efficiency versus Class

A class-A amplifier typically operates at 40% or less efficiency, depending on the nature of the input signal and the type of bipolar or field-effect transistor used. A class-AB$_1$ RF PA exhibits efficiency on the order of 35% to 45%. The efficiency of a class-AB$_2$ RF PA can approach 60%. Class-B amplifiers are 50% to 65% efficient, and class-C RF PA systems can provide up to 80% efficiency.

In extreme weak-signal class-A situations, efficiency doesn't concern us. Instead, we must focus our attention on obtaining high gain and a low *noise figure* (internally generated circuit noise), thereby optimizing the *sensitivity* of the system. We should also ensure that the circuit operates in a linear fashion.

Drive and Overdrive

In theory, class-A and class-AB$_1$ power amplifiers draw no power from a signal source to generate useful signal power output. Class-AB$_2$ amplifiers require a certain minimum amount of drive to produce signal output. Class-B amplifiers require more drive than class-AB$_2$, and class-C amplifiers need still more. Whatever class of amplification we employ, we must ensure that the driving signal doesn't get too strong. If we apply too much AC signal input to an amplifier, we create a state of *overdrive*.

Figure 7-7 illustrates output signal waveforms for power amplifiers in various situations, assuming that the input signal constitutes a perfect AC sine wave. The waveform at A shows the signal output from a properly driven class-A amplifier, and the waveform at B shows the signal output from an overdriven class-A amplifier. Drawings C and D show the output signal waveforms from properly driven and overdriven class-B RF amplifiers, respectively. Drawings E and F show the output waveforms from properly driven and overdriven class-C RF amplifiers, respectively.

TIP   *Note the* **flat topping** *that occurs with overdrive. This phenomenon can cause excessive harmonic emission, modulation-envelope distortion, and reduced efficiency. Whenever we operate a power amplifier, we must make certain that we don't overdrive it.*

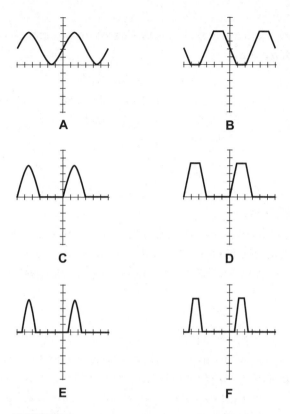

FIGURE 7-7 • Output waveforms of power amplifiers: properly driven class-A (graph A), overdriven class-A (graph B), properly driven class-B (graph C), overdriven class-B (graph D), properly driven class-C (graph E), and overdriven class-C (graph F).

PROBLEM 7-4

Suppose that an RF PA operates with a collector current of 800 mA and a collector voltage of 12 V. We get 4.8 W of useful RF power output. What's the efficiency of this amplifier, expressed in percent?

✔ SOLUTION

First, let's determine the DC power input. The collector current equals 0.800 A and the collector voltage equals 12 V. Therefore, we have

$$P_{in} = E_c I_c$$
$$= 12 \times 0.800$$
$$= 9.6\,W$$

The RF power output equals 4.8 W. Therefore, we can calculate the efficiency in percent as

$$eff_\% = (100P_{out}/P_{in})\%$$
$$= (100 \times 4.8/9.6)\%$$
$$= (100 \times 0.5)\%$$
$$= 50\%$$

PROBLEM 7-5

Suppose that we provide an amplifier with a sine-wave input signal, but the output signal waveform looks like the one shown in Fig. 7-7D. What can we say about the effect of this circuit on the signal waveform?

SOLUTION

The negative parts of every cycle are cut off, giving us the equivalent of half-wave rectification. In addition, the output waveform is distorted, not only because of the rectification, but because of a substantial change in the shapes of the portions of the cycle that remain. This situation indicates a likely state of overdrive.

Audio Amplification

High-fidelity (hi-fi) audio amplifiers work from a few hertz up to frequencies considerably beyond the limit of human hearing. A well-designed hi-fi amplifier can function from about 10 Hz to 100 kHz. Audio amplifiers for voice communications cover roughly 300 Hz to 3 kHz. In digital communications, audio amplifiers operate over a confined frequency range, often less than 1 kHz wide.

Frequency Response

Hi-fi amplifiers are equipped with resistance-capacitance (RC) *tone controls* that tailor the frequency response. The simplest tone control uses a single rotatable rotary or slide potentiometer. More sophisticated systems have separate controls, one for *bass* (low-frequency sound, roughly below the musical note *middle* C) and the other for treble (high-frequency sound, roughly above middle C). The most advanced systems have *graphic equalizers* that comprise multiple controls affecting the amplifier gain over various frequency spans. Figure 7-8 shows gain-versus-frequency curves for three hypothetical audio amplifiers. At A, we see a wideband, flat curve, typical of hi-fi system amplifiers. At B, we see a voice communications response. At C, we see a narrowband response curve, typical of audio amplifiers in digital wireless receivers.

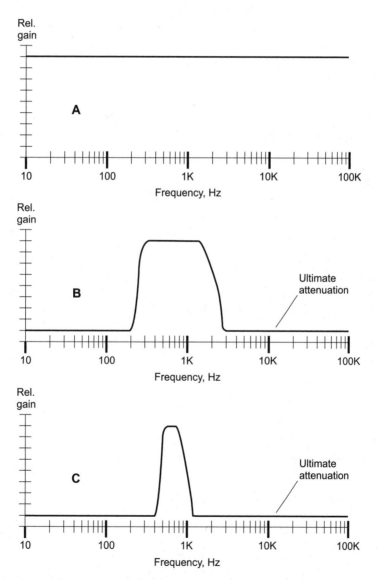

FIGURE 7-8 · Audio amplifier frequency response curves: typical hi-fi (graph A), voice communications (graph B), and digital communications (graph C).

Volume (Gain) Control

Audio amplifier systems usually have two or more stages, each with its own transistor. In an early stage with low signal power, we can include a *volume control*, such as the one shown in Fig. 7-9. In this diagram, potentiometer R_1 serves as the volume control, also called the *gain adjustment*. The AC output signal passes through capacitor C_1 and appears across R_1. The potentiometer's movable contact, known as the *wiper* and indicated by the arrow, "picks off" more or less of the AC output signal, depending on the extent to which we turn the shaft clockwise or counterclockwise. Capacitor C_2 isolates the potentiometer from the DC bias voltage of the following stage.

TIP   *Normally, we should place a volume control in a stage where the audio power level remains well below 1 W. This allows us to employ a common, low-cost potentiometer, rated at 1 or 2 W maximum.*

Coupling Methods

The term *coupling* refers to any method of getting a signal from one stage in a complex circuit to the next, such as from a low-level amplifier to a power amplifier. Figure 7-10 illustrates an example of *transformer coupling*. Capacitors C_1 and C_2 keep one end of the transformer primary and secondary at signal ground. Resistor R_1 limits the current through Q_1, the first transistor. Resistors R_2 and R_3 provide the proper base bias for transistor Q_2. We match the output impedance of Q_1 to the input impedance of Q_2 by using a transformer having the correct turns ratio.

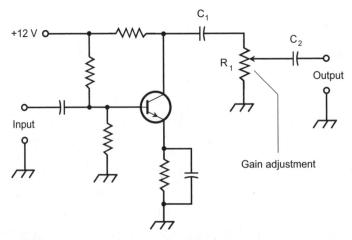

FIGURE 7-9 · A simple audio-amplifier volume (gain) control.

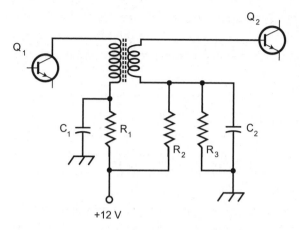

FIGURE 7-10 · An example of transformer coupling.

In some RF amplifier systems, engineers connect fixed or variable capacitors across the primary and/or secondary transformer winding to obtain *resonance* at a frequency determined by the capacitance and the winding inductance. This design method, called *tuned-circuit coupling*, can increase the efficiency of an amplifier chain. It can also improve sensitivity in weak-signal applications.

TIP *If we want to take advantage of tuned-circuit coupling, we must ensure that we don't increase the efficiency so much that the amplifier chain oscillates at the resonant frequency. If that happens, the whole amplifier system will stop working effectively.*

RF Amplification

The *RF spectrum* begins at a few kilohertz (disagreement exists in the literature about the exact lower limit) and extends upward in frequency to over 300 GHz. At the low-frequency end of this range, RF amplifiers resemble AF amplifiers. As the frequency increases, amplifier designs and characteristics change.

Weak-Signal Amplifiers

In any wireless receiver, the *front end* (the stage that gets the signal from the antenna) requires an amplifier with excellent sensitivity, which depends on the *gain* and the *noise figure*. We define and measure gain in decibels (dB). The noise figure expresses how well a circuit or system can amplify desired signals without introducing unwanted noise. All bipolar transistors or FETs generate some noise—"random" energy whose frequencies span a broad range—as charge carriers "jump" from atom to atom within the semiconductors. In general, FETs produce less noise than bipolar transistors. In the output of a receiver, noise sounds like a hiss or roar.

Weak-signal amplifiers almost always use resonant circuits. This practice optimizes the amplification at the desired frequency, while helping to minimize noise arising from energy at unwanted frequencies. Figure 7-11 shows a tuned weak-signal RF amplifier that uses a *gallium arsenide FET* (abbreviated GaAsFET and pronounced "GAS-fet"). The circuit is designed to operate at approximately 10 MHz. At higher frequencies, the inductances and capacitances would be smaller than the values shown in Fig. 7-11; at lower frequencies, the values would be larger than those shown in Fig. 7-11.

Broadband RF Power Amplifiers

A *broadband RF PA* offers comparative ease of operation because the circuit doesn't require tuning. The operator need not worry about critical adjustments. However, broadband RF PAs are slightly less efficient than their tuned counterparts.

A broadband PA amplifies any signal whose frequency lies within its design range, whether or not that signal is intended for transmission. If some circuit in

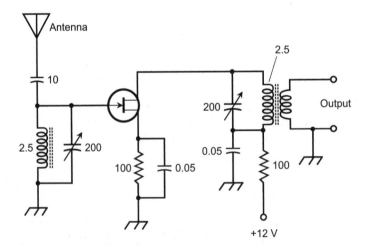

FIGURE 7-11 • A tuned weak-signal RF amplifier for operation at about 10 MHz. Resistances are in ohms. Capacitances are in microfarads (μF) if less than 1, and in picofarads (pF) if greater than 1. Inductances are in microhenrys (μH).

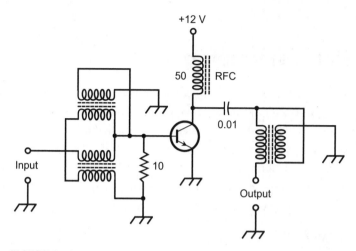

FIGURE 7-12 • A broadband RF power amplifier, capable of producing a few watts output. Resistances are in ohms. Capacitances are in microfarads (μF) if less than 1, and in picofarads (pF) if greater than 1. Inductances are in microhenrys (μH). The radio-frequency choke (RFC) blocks high-frequency AC but allows DC and low-frequency AC to pass.

a transmitter has too much feedback, causing *oscillation* at a frequency other than the intended signal frequency, and if this undesired signal falls within the design frequency range of a broadband PA, the unwanted signal will undergo amplification along with the desired signal, causing *spurious emissions*. These emissions constitute an "operational emergency" because they can interfere with other communications systems.

Figure 7-12 shows a typical broadband PA circuit using an NPN power transistor. The transformers form a critical part of this circuit. They must function efficiently over a wide range of frequencies. This circuit will work well from approximately 1 MHz through 30 MHz.

**PROBLEM 7-6**

What do the dashed lines represent in the transformer and inductor symbols in Figs. 7-10, 7-11, and 7-12?

SOLUTION

Dashed lines between the windings of a transformer, or adjacent to the windings of an inductor, represent powdered-iron cores.

Still Struggling

You might ask why an engineer would want to use powdered-iron cores in the circuits of Figs. 7-10, 7-11, and 7-12, rather than solid or laminated iron cores (which would show up as solid lines, rather than dashed lines, in the transformer symbols). Solid or laminated iron transformer and inductor cores don't work well in RF applications because circulating currents arise within them, heating them up and causing the transformers or inductors to waste otherwise useful RF signal power.

PROBLEM 7-7

Why are powdered-iron cores specified in the diagrams of Figs. 7-10, 7-11, and 7-12, rather than air cores?

SOLUTION

When a transformer or inductor has a powdered-iron core, the inductance of each coil winding increases compared to the inductance we would get with an air core, assuming the same number of turns and the same coil dimensions. Stated another way, powdered-iron cores allow us to employ transformers and coils having fewer turns than we would need if the cores were air. In addition, powdered iron concentrates most of the magnetic flux in the core material, so relatively little of it appears outside the coil where it might disrupt the operation of other components. Air-core transformers and coils can be (and sometimes are) used in RF applications, where the required inductances are fairly small. Air-core coils don't work well in AF applications because they don't allow us to obtain enough inductance.

Tuned RF Power Amplifiers

A *tuned RF PA* offers improved efficiency compared with broadband designs. Also, the tuning helps to reduce the risk that spurious signals will be amplified and transmitted over the air. A tuned RF PA can deliver output into circuits or devices having a wide range of impedances. In addition to a *tuning control*, or resonant circuit that adjusts the output of the amplifier to the operating frequency, a *loading control* optimizes the signal transfer between the amplifier and the load.

The main drawbacks of a tuned PA are the facts that adjustment can consume a lot of operator time, and improper adjustment can destroy the amplifying

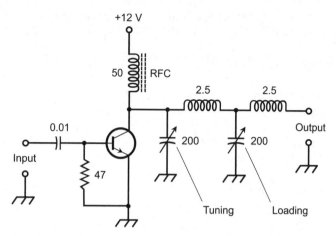

FIGURE 7-13 · A tuned RF power amplifier, capable of producing a few watts of useful power output. Resistances are in ohms. Capacitances are in microfarads (μF) if less than 1, and in picofarads (pF) if greater than 1. Inductances are in microhenrys (μH).

device (bipolar transistor or FET). If we don't set the tuning and/or loading controls correctly, the efficiency will drop to near zero while the DC collector or drain power input remains high. In fact, the DC power input might actually increase compared with the situation in which the controls are properly adjusted. As a result, the amplifying device can overheat and fail.

Figure 7-13 is a schematic diagram of a tuned RF PA, providing useful power output at approximately 10 MHz. We must adjust the "tuning" and "loading" controls (variable capacitors having maximum values of 200 pF in this case) for maximum RF power output, as indicated on a wattmeter connected to the output terminals.

QUIZ

Refer to the text in this chapter if necessary. A good score is eight correct. The answers are listed in the back of the book.

1. Which of the following RF power-amplification modes will work well for FM signals but will likely cause distortion in AM signals?
   A. Class A
   B. Class AB$_1$ or AB$_2$
   C. Class B
   D. Class C

2. **If we set the tuning and/or loading controls improperly in a tuned bipola transistor RF amplifier, which of the following effects, if any, will we likely observe?**

   A. Excessive heat in the transistor
   B. Excessive harmonic output
   C. Excessive signal power output
   D. All of the above

3. **Which of the following amplifier types exhibits the best linearity for the signal waveform?**

   A. Class A
   B. Class AB_1 or class AB_2
   C. Class B
   D. Class C

4. **Figure 7-14 illustrates a generic broadband audio amplifier using an op amp. What, if anything, is wrong with this circuit?**

   A. The potentiometer should go from the output to the noninverting input.
   B. The potentiometer should go between the inverting and noninverting inputs.
   C. We should replace the potentiometer with a variable capacitor.
   D. Nothing is wrong with this circuit as shown.

5. **Suppose that we adjust the potentiometer in the circuit of Fig. 7-14 to reduce its resistance. What can we expect as a result?**

   A. The circuit will more likely oscillate.
   B. The circuit gain will decrease.
   C. The frequency response will get more narrow.
   D. All of the above

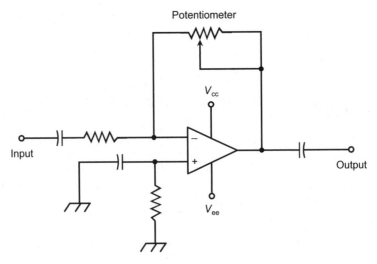

FIGURE 7-14 · Illustration for Quiz Questions 4 and 5.

6. If we *increase* the RMS voltage of an AC signal by a factor of 20 without altering the impedance in any way, then we introduce an amplitude change of
   A. +10 dB.
   B. +13 dB.
   C. +20 dB.
   D. +26 dB.

7. If we *decrease* the RMS power of an AC signal by a factor of 20, then we introduce a power gain of
   A. −10 dB.
   B. −13 dB.
   C. −20 dB.
   D. −26 dB.

8. Suppose that we employ a transformer to couple the output of a weak-signal JFET RF amplifier to the input of a bipolar-transistor amplifier that boosts the signal strength further. We place capacitors across the transformer windings to obtain resonance at the received-signal frequency. What's the principal disadvantage of this arrangement?
   A. The system will exhibit an excessively broad bandwidth unless we control the gain of the second amplifier.
   B. The system will likely generate too much internal noise unless we optimize the gain of the second amplifier.
   C. The system might break into oscillation if the amplifier stages exhibit too much gain.
   D. The system will generate excessive harmonic energy unless we optimize the gain of the first amplifier.

9. In which of the following amplifier types do we observe DC power input *only* when we apply an input signal?
   A. Class B
   B. Class AB_2
   C. Class AB_1
   D. Class A

10. Suppose that we want to build an amplifier to install between a microphone and an audio hi-fi system, with the intent of getting the microphone to detect faint sounds. Which of the following circuits would work best for this purpose?
    A. A class-C op amp
    B. A class-B amplifier using an N-channel JFET
    C. A class-A amplifier using a P-channel JFET
    D. A tuned class-AB_2 amplifier using an NPN transistor

chapter **8**

Signal Oscillators

An *oscillator* is an amplifier circuit with positive feedback that makes it generate a signal. Most RF oscillators generate near-perfect sine waves. In AF applications, oscillators can emit sine, square, sawtooth, or complex waves.

CHAPTER OBJECTIVES

In this chapter you will

- Learn how various forms of feedback can cause oscillation.
- Analyze RF oscillator circuits that use resonant circuits to determine their frequencies.
- Learn how to enhance the stability and reliability of an oscillator.
- See how crystal-controlled oscillators work.
- Discover the concept of phase locking to enhance oscillator stability.
- Compare twin-T and multivibrator AF oscillator designs.

Oscillators for RF

In order for a circuit to oscillate, it must have high gain and positive feedback. Common-emitter and common-source circuits work well as the basis for oscillator design. Common-base and common-gate circuits can oscillate under certain conditions, but they more often find application as power amplifiers.

Feedback

We can control the frequency of an RF oscillator by means of a tuned, resonant circuit in the *feedback loop* as shown in Fig. 8-1. This resonant circuit usually comprises an inductance-capacitance (*LC*) or resistance-capacitance (*RC*) combination. Engineers prefer to use the *LC* scheme in RF oscillators and the *RC* method in AF oscillators.

The tuned circuit in an oscillator exhibits its minimum loss at a single frequency, and higher loss at all other frequencies. As a result, oscillation occurs at a predictable and stable frequency determined by the inductance and capacitance, or by the resistance and capacitance.

Armstrong Circuit

A common-emitter or common-source RF amplifier will oscillate if we feed some of the output signal back to the input, in phase, through a transformer. The signal wave at the collector or drain exists "upside-down" with respect to the signal wave at the base or gate, so we must deliberately invert the wave again to obtain positive feedback. We can accomplish this task by connecting the transformer secondary to the base or gate "backwards."

Figure 8-2A shows a common-emitter NPN bipolar-transistor RF amplifier in which we couple the collector circuit to the emitter circuit through a

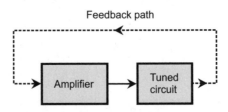

FIGURE 8-1 · An oscillator works by feeding some of the output signal back to the input through a tuned circuit that determines the frequency.

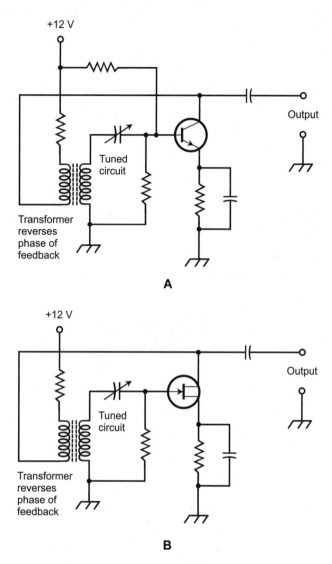

FIGURE 8-2 · Armstrong oscillators. At A, we use an NPN bipolar transistor. At B, we use an N-channel JFET.

transformer that reverses the phase of the feedback signal. Figure 8-2B illustrates the equivalent N-channel JFET circuit. Engineers call circuits of this type *Armstrong oscillators*.

We can adjust the oscillation frequency by installing a variable capacitor in either the primary or the secondary circuit of the feedback transformer. The inductance of the winding, along with the capacitance, forms an RF resonant circuit. We can calculate the fundamental oscillation frequency with the formula

$$f = 1 / [2\pi(LC)^{1/2}]$$

where f represents the frequency in megahertz, L represents the inductance of the transformer winding in microhenrys, and C represents the value of the variable capacitor in microfarads.

Hartley Circuit

Figure 8-3 illustrates a second method of obtaining controlled feedback. The circuit at A employs an NPN bipolar transistor; the circuit at B employs an N-channel JFET. In both cases, the circuit has a coil with a tap to provide the

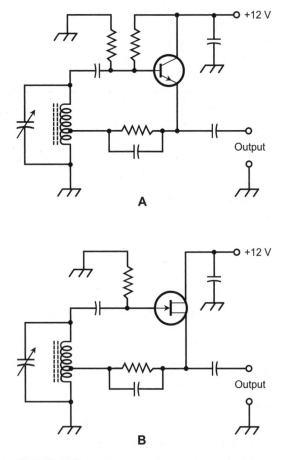

FIGURE 8-3 · Hartley oscillators. At A, we use an NPN bipolar transistor. At B, we use an N-channel JFET.

feedback. A variable capacitor in parallel with the coil determines the oscillating frequency. The formula for resonant frequency applies, just as in the Armstrong case. A circuit of this type constitutes a *Hartley oscillator*.

A properly designed Hartley oscillator feeds about 25% of its output signal power back to the input circuit to sustain the oscillation. The remaining 75% of the power can provide useful signal output. Oscillators rarely produce more than a fraction of a watt of useful output power. If we want to get more power than that, we must "boost" the signal with at least one amplifier circuit following the oscillator.

TIP   *When we design and build a Hartley oscillator, we should use the minimum amount of feedback necessary to obtain oscillation. The amount of feedback depends on the position of the coil tap. As we move the coil tap toward the ground end and away from the base or gate end, we decrease the feedback; as we move the tap away from the ground end and toward the base or gate end, we increase the feedback. In any event, the signal feedback occurs between the emitter and base (for a bipolar transistor) or between the source and gate (for an FET).*

Colpitts Circuit

Another way to provide RF feedback involves tapping the capacitance instead of the inductance in a parallel-tuned resonant *LC* circuit. Figure 8-4 shows NPN bipolar (at A) and N-channel JFET (at B) *Colpitts oscillator* circuits.

In the Colpitts system, we control the amount of feedback by adjusting the ratio of the series capacitances that appear across the coil. The oscillation frequency depends on the net series capacitance C, and also on the inductance *L* of the variable coil. The general formula for resonant frequency applies, just as it does for the Armstrong and Hartley oscillators.

If C_1 and C_2 represent the values of the two series capacitors connected across the variable inductor, then we can calculate the net capacitance C using the formula

$$C = C_1 C_2 / (C_1 + C_2)$$

If we want this formula to work, we must express all capacitance values in the same units (usually picofarads in an RF oscillator).

In the Colpitts circuit, we'll find it more convenient to vary the inductance, rather than the capacitance, to adjust the oscillation frequency. We can vary the inductance of a solenoidal (cylindrical) coil by sliding a powdered-iron rod called a *slug* in and out of a coil wound on a hollow plastic or ceramic form. A properly designed Colpitts circuit offers exceptional stability and reliability.

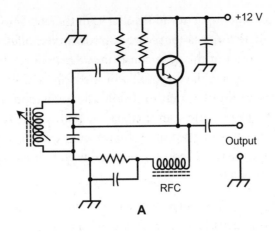

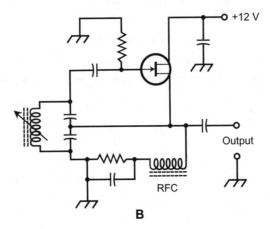

FIGURE 8-4 · Colpitts oscillators. At A, we use an NPN bipolar transistor. At B, we use an N-channel JFET.

PROBLEM 8-1

In the Colpitts oscillator circuit of Fig. 8-4B, what's the purpose of the fixed capacitor between the tuned resonant *LC* circuit and the gate of the JFET?

SOLUTION

This capacitor allows us to set the DC bias at the gate by connecting a resistor between the gate and ground. If the fixed capacitor were not there, the gate would be shorted to ground for DC through the coil in the tuned circuit, forcing the JFET to operate in a zero-bias condition.

TIP *As we would do in a Hartley circuit, we should keep the feedback in a Colpitts oscillator to the minimum necessary to sustain oscillation. The amount of feed-back depends on the ratio of the capacitances across the variable inductor. As we make the base-side or gate-side capacitance smaller and the ground-side capacitance larger, we reduce the feedback. As we make the base-side or gate-side capacitance larger and the ground-side capacitance smaller, we increase the feedback. As with the Hartley circuit, feedback occurs between the emitter and base (for a bipolar transistor) or between the source and gate (for an FET).*

Still Struggling

Do you wonder why, in the Hartley and Colpitts circuits shown in Figs. 8-3 and 8-4, we take the output from the emitter or source circuit, rather than from the collector or drain circuit? As a matter of fact, we can take the output from the collector or drain, and the circuit will probably work "after a fashion." However, if we do that, we increase the risk of ending up with an oscillator with poor *stability* characteristics. In particular, the oscillator might prove unreliable. We'll define and examine stability factors shortly.

Clapp Circuit

A variation of the Colpitts oscillator employs a series-resonant tuned circuit. Figure 8-5A is a schematic diagram of an NPN bipolar-transistor *Clapp oscillator*. Figure 8-5B shows the equivalent N-channel JFET circuit.

The Clapp oscillator design offers excellent stability in RF applications. If we use high-quality components (particularly the inductor and the capacitors in the resonant circuit), the Clapp circuit's frequency will fluctuate, or *drift*, less under varying ambient conditions than will an equivalent Hartley or Colpitts circuit's frequency. We can easily get a Clapp circuit to start oscillating when we power it up, and it will keep oscillating in situations where other designs would fail. We can use a variable capacitor to control the frequency, an advantage over the conventional Colpitts design because variable capacitors cost less than variable inductors.

To prevent the output signal from shorting to ground, we connect an *RF choke* (RFC) in series with the emitter or source in the Colpitts and Clapp

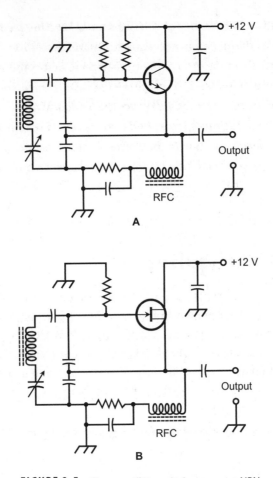

FIGURE 8-5 · Clapp oscillators. At A, we use an NPN bipolar transistor. At B, we use an N-channel JFET.

circuits. Typical inductance values for RF chokes are 100 μH at high frequencies such as 15 MHz, and 10 mH at low frequencies such as 150 kHz. As the term implies, an RF choke blocks RF currents while allowing DC and low-frequency AC to pass through unimpeded.

PROBLEM 8-2

In the Clapp oscillator circuit of Fig. 8-5A, how does the oscillation frequency relate to the value of the variable capacitor in picofarads?

✔ SOLUTION

As you increase the value of the variable capacitor in picofarads, the capacitance of the set of capacitors (four of them exist in series as you go around the resonant circuit) also increases. When you scrutinize the formula for the resonant frequency of a tuned circuit, you'll see that the frequency varies inversely in proportion to the square root of the capacitance. Therefore, as the capacitance of the variable capacitor increases, the oscillation frequency goes down. As the capacitance of the variable capacitor decreases, the oscillation frequency goes up.

TIP *At ultra-high frequencies (UHF) and microwave radio frequencies, specialized semiconductor diodes can function as oscillators. Examples include Gunn, IMPATT, and tunnel diodes. You learned about them in Chapter 5.*

Oscillator Stability

With respect to oscillators, the term *stability* can refer to constancy of frequency or general reliability.

Frequency Stability

Fluctuations in the ambient temperature commonly affect electronic component values. Whenever we build an oscillator, we should take precautions to ensure that the values of the components, especially the inductors and capacitors, vary to the smallest extent possible (except when we deliberately adjust a capacitor or inductor to control the frequency).

Some types of capacitors maintain their values better than others as the temperature rises or falls. *Polystyrene capacitors* behave very well in this respect. *Silver-mica capacitors* do a decent job when we can't find polystyrene capacitors. *Air-variable capacitors* exhibit the best stability of all, although we should protect them from severe physical vibration or shock that might upset their settings.

Inductors are most temperature-stable when their cores comprise only air. The best air-core coils have stiff wire supported by strips of rigid plastic to keep the windings in place. Some air-core coils have hollow forms made of ceramic or phenolic material, but the wall thickness should be kept to a minimum.

Ferromagnetic core materials such as powdered iron can function in some RF applications, but temperature changes affect the *core permeability*. Any variation in core permeability alters the inductance, in turn affecting the frequency of an oscillator that uses the coil in its resonant circuit.

TIP *Problems with oscillator frequency stability have been largely overcome in recent years by the evolution and proliferation of* frequency synthesizers *in RF systems. We'll learn how they work later in this chapter.*

Reliability

An oscillator should begin functioning as soon as we apply power, and it should keep oscillating under all normal conditions, not quitting if the load changes or the temperature suddenly rises or falls. The failure of a single oscillator can cause an entire communications system to go down!

A typical oscillator "prefers" to operate with a high *load impedance*, meaning that the circuit connected to the oscillator output should exhibit a high, purely resistive input impedance. If we connect a low-impedance device or system to the output of an oscillator, the load will "try" to draw power from the oscillator. Under these conditions, even a well-designed oscillator can become unreliable. When we need useful signal power, we should obtain it by placing one or more amplifiers after the oscillator. We shouldn't try to make an oscillator provide significant power all by itself.

When we build a *variable-frequency oscillator* (VFO) and place it in service, we should expect to go through a *debugging* process. If we construct two VFOs from the same schematic diagram, with the same component types and in the same physical arrangement, one circuit might exhibit good stability while the other drifts or intermittently fails. When we encounter this sort of situation, we can often trace the trouble to a defect in a single component. However, in some complicated systems, problems arise that defy debugging.

Still Struggling

Some engineers and technicians, when frustrated by a recalcitrant problem, have been known to blame *gremlins*—tiny, imaginary monsters—for the malfunction. According to the mythology, these creatures exist for the sole purpose of wreaking havoc in electronic, computer, and mechanical systems. If anyone ever tells you that they've actually seen a gremlin, refuse to believe them. Gremlins play their games only when nobody's looking!

Crystal-Controlled Oscillators

If we don't often need to change the frequency of an RF oscillator, we can use a *quartz crystal*, also called a *piezoelectric crystal*, in place of a resonant *LC* circuit. *Crystal-controlled oscillators* offer frequency stability superior to that of *LC*-tuned oscillators.

Pierce Circuit

Figure 8-6 illustrates a *Pierce oscillator*, one of the most common crystal-controlled designs. At A, we use an NPN bipolar transistor. At B, we use an

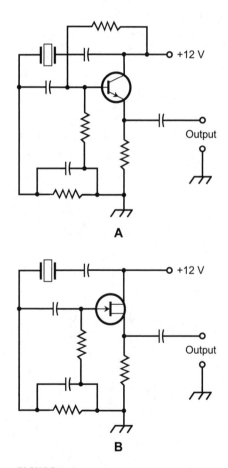

FIGURE 8-6 · Pierce oscillators. At A, we use an NPN bipolar transistor. At B, we use an N-channel JFET.

N-channel JFET. We can *trim* (adjust) the frequency by approximately ±0.1% of the design frequency if we connect an inductor in series with the crystal, or if we connect a capacitor in parallel with the crystal.

In any crystal-controlled oscillator, the fundamental oscillation frequency depends on the thickness of the crystal, and also on the angle at which the crystal was originally cut from the quartz specimen. Crystals change in frequency as the temperature changes. However, in most practical situations, crystals exhibit better frequency stability than resonant *LC* circuits do.

TIP   *Some crystal oscillators reside in thermally insulated, temperature-controlled chambers called* **crystal ovens.** *These crystals maintain their frequency so precisely that we can use them as* **frequency standards** *to calibrate other oscillators. The frequency rarely departs by more than a few* **parts per million** *(for example, a few hertz at working frequencies of several hundred kilohertz).*

Reinartz Circuit

A *Reinartz crystal oscillator* offers high efficiency and minimal harmonic signal output. As with other oscillators, we can use either a bipolar transistor or an FET as the active component. Figure 8-7 is a schematic diagram of a Reinartz oscillator that employs an N-channel JFET. We insert a parallel-resonant *LC* circuit in the source line and tune it to approximately half the crystal frequency. This design scheme causes the circuit to oscillate at a low level of crystal current. We tune the parallel-resonant *LC* circuit in the drain line, formed by the

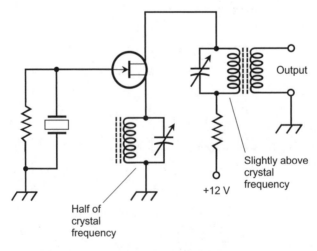

FIGURE 8-7 · A Reinartz crystal oscillator using an N-channel JFET.

combination of a variable capacitor and the output-transformer primary, to a frequency slightly higher than the crystal's fundamental design frequency.

Variable-Frequency Crystal Oscillator (VXO)

The frequency of a *variable-frequency crystal oscillator* (VXO) can be varied slightly with the addition of a reactance in series or parallel with the crystal. Engineers refer to this practice as *frequency trimming.* VXOs can function in RF transmitters or transceivers to obtain operation over a narrow range of frequencies. Figure 8-8 shows two examples of bipolar-transistor VXO circuits. At A,

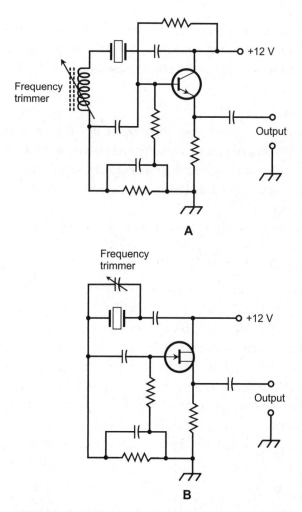

FIGURE 8-8 · Bipolar-transistor VXO circuits. At A, inductive frequency trimming; at B, capacitive frequency trimming.

we connect a variable inductor in series with the crystal. At B, we connect a variable capacitor in parallel with the crystal. The main advantage of the VXO over an ordinary VFO is enhanced frequency stability. The main disadvantage of the VXO is limited frequency coverage.

TIP *In most "real-world" RF systems, we cannot vary the frequency of a VXO by more than one part per thousand, or approximately ±0.1% of the operating frequency, without loss of reliability. At 10 MHz, for example, that constitutes a frequency-trimming range of only about ±10 kHz.*

PROBLEM 8-3

In the crystal-controlled oscillator circuit shown in Fig. 8-8B, what would happen if the resistor between the gate and ground were to fail, creating a short circuit from the gate directly to ground?

SOLUTION

This malfunction would place the gate at signal ground as well as at DC ground, shorting the input signal to the transistor and altering the gate bias. The JFET acts as an amplifying device. If we deprive it of an input signal, we'll obviously keep it from producing any output. The oscillator would stop working.

Voltage-Controlled Oscillator (VCO)

We can adjust the frequency of a VFO by means of a varactor diode in the tuned *LC* circuit. Hartley and Clapp oscillator circuits lend themselves to varactor frequency control. We use a blocking capacitor to isolate the varactor with respect to DC, and then apply a variable DC voltage directly to the varactor. Figure 8-9 shows a JFET Hartley *voltage-controlled oscillator* (VCO) in which a varactor diode and a fixed capacitor replace the variable capacitor. The capacitor in series with the varactor keeps the positive control voltage from shorting to ground through the coil. (The negative power-supply terminal goes directly to ground by design; we call this arrangement a *negative-ground system*.)

PROBLEM 8-4

In the oscillator circuit shown in Fig. 8-9, what would happen if the capacitor between the source and the output terminal were to open?

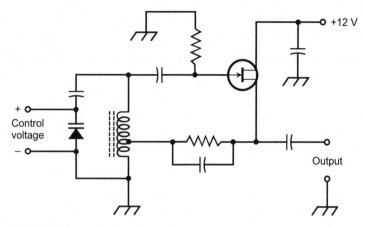

FIGURE 8-9 · A Hartley voltage-controlled oscillator (VCO) using an
N-channel JFET.

 **SOLUTION**

**This action wouldn't stop the circuit from oscillating, but it would prevent
the signal from reaching the output terminal. Therefore, the effects on circuits following the oscillator would mimic an oscillator failure.**

PLL Frequency Synthesizer

A *phase-locked-loop (PLL) frequency synthesizer* combines the flexibility of a
VFO with the stability of a crystal oscillator. In recent decades, *frequency syn-
thesis* has gained popularity in RF wireless equipment of all kinds.

Figure 8-10 shows the basic components of a PLL frequency synthesizer. The
output signal from a VCO passes through a *programmable divider*, a digital

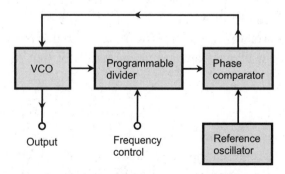

FIGURE 8-10 · Block diagram of a phase-locked-loop
(PLL) frequency synthesizer.

circuit that can divide the VCO frequency by any *integral* (whole-number) factor we want. A *phase comparator* locks the output frequency of the programmable divider to the frequency of the signal from a crystal-controlled *reference oscillator*. As long as the output from the programmable divider remains exactly on the reference-oscillator frequency, the two signals stay in phase coincidence, and the output of the comparator equals zero volts (0 V DC). If the VCO output frequency "tries" to drift, the output frequency of the programmable divider also "tries" to drift, although at a different rate.

The slightest change in the VCO frequency causes the phase comparator to produce a *DC error voltage*. This voltage has either positive or negative polarity, depending on whether the VCO has drifted higher or lower in frequency. We apply the DC error voltage to a varactor in the VCO, causing the VCO frequency to change in a direction opposite that of the drift. This action creates a *DC feedback loop* that maintains the VCO frequency at a precise integral ratio relative to the reference-oscillator frequency (that ratio having been chosen by the programmable divider).

TIP   *Engineers can enhance the frequency stability of a PLL by using, as the reference oscillator, an amplified signal from the National Institute of Standards and Technology (NIST). These signals are transmitted on the "shortwave radio" bands at 5, 10, or 15 MHz, and also on the "longwave radio" band at 60 kHz. Do you have one of those fancy "atomic-standard" clocks that keeps practically perfect time? It uses one of the NIST signals to keep its internal oscillator locked on frequency.*

Audio Oscillators

Audio-frequency (AF) oscillators find application in doorbells, ambulance sirens, electronic games, personal computers, and toys that play musical tunes. All audio oscillators consist of AF amplifiers with positive feedback. Most audio oscillators use resistance-capacitance (*RC*) circuits to determine the frequency. You might occasionally encounter an audio oscillator that uses an *LC* circuit —but not often!

Waveforms

In RF applications, most oscillators are designed to produce a nearly perfect sine-wave output, representing energy at a single, well-defined frequency. But audio oscillators don't necessarily concentrate all their energy at one frequency. Applications exist, particularly in electronic music, where an end user wants almost anything *but* a sine wave.

Various musical instruments sound different, even when they play notes at the same pitch. We hear this difference because each instrument has its own unique waveform. An instrument's essential sound quality (or *timbre*) can be reproduced using an audio oscillator whose waveform output matches that of the instrument. The waveform depends on the relative energy content in the harmonics (multiples of the fundamental frequency).

TIP   *A computer can have a* **musical-instrument digital interface** *(MIDI) player that employs audio oscillators capable of duplicating the sounds of a large band or orchestra. This arrangement allows you to produce sophisticated electronic music using a modest personal computer. A typical MIDI system can generate almost any imaginable audio waveform, creating notes that sound like those of any musical instrument you want.*

Twin-T Oscillator

We can build a popular general-purpose circuit for generating AF signals in the form of a *twin-T oscillator* (Fig. 8-11). The transistor on the right provides

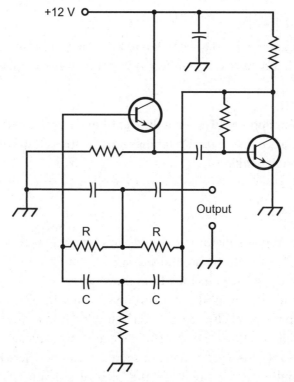

FIGURE 8-11 · A twin-T audio oscillator.

amplification that keeps the oscillation going. The two resistors R and the two capacitors C determine the fundamental oscillation frequency, and also ensure that the feedback occurs in phase coincidence. The transistor on the left acts as an emitter follower, stabilizing the circuit so that its oscillation frequency doesn't wildly fluctuate (as it can do in poorly designed audio oscillators). We "pick off" the output from between the two resistors marked R. The circuit in this example uses NPN bipolar transistors. Similar circuits can, of course, be built using N-channel JFETs, PNP bipolar transistors, or P-channel JFETs.

PROBLEM 8-5

In the circuit of Fig. 8-11, what would happen if the capacitor C on the left-hand side were to short out?

SOLUTION

This malfunction would deprive the left-hand transistor of its input signal, interrupting the feedback loop and causing the oscillator to stop working.

PROBLEM 8-6

In the circuit of Fig. 8-11, what would happen if the connection between the right-hand transistor's emitter and ground were to open up?

SOLUTION

This malfunction would prevent the right-hand transistor from amplifying, causing a lack of feedback. Under these circumstances, the oscillator would probably quit working.

Multivibrator

A *multivibrator* type oscillator circuit employs two amplifier stages, interconnected so that the signal goes around and around between them. We can build a multivibrator using N-channel JFETs, as shown in Fig. 8-12. Each JFET amplifies the signal in class-A mode, and also inverts the signal waveform. In this particular circuit, a parallel-resonant *LC* circuit determines the frequency.

If we want the circuit of Fig. 8-12 to generate an AF signal, we must use a large-value inductor. Ideally, the inductor should have a powdered-iron core that surrounds a multiturn wire loop like the "skin of a donut." This core shape,

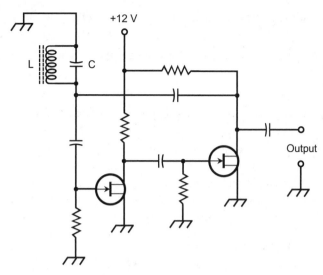

FIGURE 8-12 · A multivibrator audio oscillator.

known as a *pot core*, allows for a much higher inductance, for a given number of coil turns, than a solenoidal (rod-shaped) core. The value of the inductor can range between approximately 0.5 H and several henrys. The large inductor allows us to use a small-value capacitor, or even an air-variable capacitor, to obtain an audio note at any desired frequency. It's an "old-school" approach to AF oscillation dating back to the 1960s, but it works today just as well as it did then!

QUIZ

Refer to the text in this chapter if necessary. A good score is eight correct. The answers are listed in the back of the book.

1. **If we want to make an amplifier function as an oscillator, we can usually do it by**
   A. introducing negative feedback.
   B. reducing the gain.
   C. operating it in the class-C mode.
   D. None of the above

2. **We can recognize a Pierce oscillator circuit by the presence of a**
   A. split capacitance in a tuned *LC* circuit.
   B. tapped inductance in a tuned *LC* circuit.
   C. piezoelectric crystal.
   D. potentiometer in parallel with a tapped inductance.

3. **Figure 8-13 is meant to show a schematic diagram of a**
   A. Hartley VCO.
   B. Colpitts VXO.
   C. Pierce PLL.
   D. multivibrator.

4. **What, if anything, is wrong with the circuit shown in Fig. 8-13?**
   A. Nothing is wrong with this circuit.
   B. We need to insert a fixed capacitor between the positive power supply terminal and the varactor cathode.
   C. We need to insert a variable capacitor in parallel with the coil.
   D. We need to insert a fixed capacitor between the varactor cathode and the top of the coil.

5. **In a PLL, a phase comparator introduces a DC error voltage when**
   A. the output signal amplitude falls below a certain threshold determined by the phase comparator.
   B. the output signal waveform departs significantly from a pure sinusoid as produced by mixing between the VCO signal and the reference-oscillator signal.
   C. the programmable divider's output frequency "tries" to depart from the reference oscillator's output frequency.
   D. the reference oscillator signal contains excessive harmonic energy, confusing the phase comparator.

6. **We can recognize a Hartley oscillator circuit by the presence of a**
   A. split capacitance in a tuned *LC* circuit.
   B. tapped inductance in a tuned *LC* circuit.
   C. piezoelectric crystal.
   D. potentiometer in parallel with a tapped inductance.

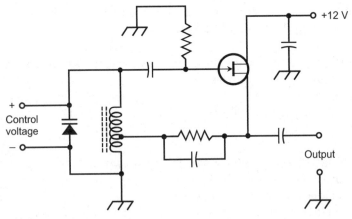

FIGURE 8-13 · Illustration for Quiz Questions 3 and 4.

7. **We can recognize a VCO by the presence of**
    A. a quartz crystal.
    B. a tapped inductor.
    C. a varactor diode.
    D. an *RC* resonant circuit.

8. **Which of the following characteristics represents an advantage of a Clapp oscillator over a conventional Colpitts oscillator?**
    A. We can use a tapped inductor to control the frequency.
    B. We can use a variable capacitor to control the frequency.
    C. We can use a potentiometer to control the frequency.
    D. We can use a PLL to control the frequency.

9. **Which of the following characteristics constitutes a principal advantage of a Pierce oscillator over a Hartley oscillator?**
    A. The Pierce circuit has better frequency stability.
    B. We can more easily vary the frequency in the Pierce circuit.
    C. The Pierce circuit has more harmonic content.
    D. The Pierce circuit generates a better sine wave.

10. **We can recognize a Colpitts oscillator in a schematic diagram by the presence of a**
    A. split capacitance in a tuned *LC* circuit.
    B. tapped inductance in a tuned *LC* circuit.
    C. piezoelectric crystal.
    D. potentiometer in parallel with a tapped inductance.

Test: Part II

Do not refer to the text when taking this test. You may draw diagrams or use a calculator if necessary. A good score is at least 38 correct. Answers are in the back of the book. It's best to have a friend check your score the first time, so you won't memorize the answers if you want to take the test again.

1. Which of the following characteristics makes a varactor diode useful for adjusting the frequency of an oscillator?

   A. High avalanche voltage
   B. Low forward-breakover voltage
   C. Variable reverse-bias capacitance
   D. Excellent linearity
   E. Efficient harmonic generation

2. In an amplifier using a P-channel JFET, we define the DC power input as the

   A. ratio of the drain current to the drain voltage.
   B. product of the drain current and the drain voltage.
   C. ratio of the drain current to the gate current.
   D. ratio of the drain current to the source current.
   E. drain voltage times the sum of the source, gate, and drain currents.

3. If we connect the negative terminal of a battery to the N type side of a semiconductor diode, and then connect the positive battery terminal to the P type side of the diode through a current-limiting resistor, the diode's P-N junction experiences

   A. saturation.
   B. forward bias.
   C. cutoff.
   D. reverse bias.
   E. avalanche.

4. In the feedback loop of an op-amp circuit, large resistances between the inverting output and the input produce high gain by

   A. minimizing the effects of external reactance.
   B. increasing the gate bias.
   C. minimizing the negative feedback.
   D. narrowing the bandwidth.
   E. placing the internal transistors beyond cutoff.

5. In an NPN bipolar transistor, the emitter consists of

   A. a P-N junction.
   B. a channel.
   C. N type material.
   D. a substrate.
   E. intrinsic semiconductor material.

6. In the circuit of Fig. Test II-1, the components marked X are

   A. varactors.
   B. PIN diodes.
   C. Gunn diodes.
   D. point-contact diodes.
   E. thyrectors.

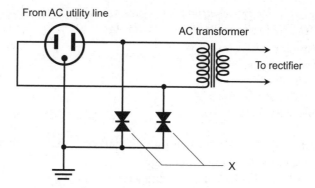

FIGURE TEST II-1 · Illustration for Part II Test Questions 6 and 7.

7. In the circuit of Fig. Test II-1, the components marked X serve to
   A. suppress transients.
   B. limit the current.
   C. regulate the voltage.
   D. prevent power overload.
   E. prevent avalanche effect.

8. If we want to build an RF power amplifier and we don't care whether it's linear or not, we can obtain the highest efficiency, in theory, if we design the circuit to operate in
   A. class A.
   B. class AB_1.
   C. class AB_2.
   D. class B.
   E. class C.

9. Imagine that we want to design a mixer circuit that will allow the reception of radio signals over a continuous span from 500 kHz to 1500 kHz. What range of frequencies can a continuously tunable local oscillator (LO) have if we want the mixer to produce an output signal at a constant frequency of 9.000 MHz?
   A. 9.500 MHz to 10.500 MHz
   B. 8.500 MHz to 10.500 MHz
   C. 8.500 MHz to 9.500 MHz
   D. 7.500 MHz to 8.500 MHz
   E. More than one of the above

10. The sensitivity of a weak-signal amplifier depends primarily on two factors. Which two?
    A. Gain and noise figure
    B. Efficiency and gain
    C. DC power input and efficiency
    D. Linearity and gain
    E. Linearity and efficiency

11. **A common-source JFET amplifier circuit that uses positive feedback through a transformer to generate an RF signal constitutes**

    A. a multivibrator.
    B. a linear power amplifier.
    C. a balanced mixer.
    D. an Armstrong oscillator.
    E. a phase-locked loop.

12. **In the circuit of Fig. Test II-2, the component marked X should exhibit**

    A. the highest possible avalanche voltage.
    B. the smallest possible junction capacitance.
    C. the largest possible current-carrying capacity.
    D. the largest possible forward-breakover voltage.
    E. All of the above

13. **In the circuit of Fig. Test II-2, the component marked X serves as**

    A. a harmonic generator.
    B. a mixer.
    C. an envelope detector.
    D. a switch.
    E. a surge protector.

14. **In a semiconductor diode, the N type wafer constitutes the**

    A. emitter.
    B. substrate.
    C. anode.
    D. cathode.
    E. collector.

15. **Which of the following terms might an engineer correctly use to describe the dynamic current gain of a PNP transistor in the grounded-base configuration?**

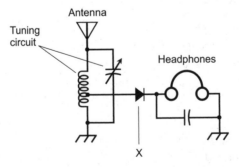

FIGURE TEST II-2 · Illustration for Part II Test Questions 12 and 13.

A. Beta
B. Alpha
C. Gamma
D. Alpha cutoff
E. Gamma cutoff

16. **Figure Test II-3 illustrates the basic configuration for**
A. a weak-signal amplifier.
B. an RF oscillator.
C. a common-source circuit.
D. an emitter follower.
E. a common-base circuit.

17. **What function does the resistor marked R serve in the circuit of Fig. Test II-3?**
A. It keeps the drain from going into a state of saturation.
B. It prevents excessive current from flowing in the gate circuit.
C. It ensures that the base remains at signal ground.
D. It keeps the input signal from shorting to ground.
E. It ensures that the emitter-base junction remains reverse-biased.

18. **What function does the capacitor marked C serve in the circuit of Fig. Test II-3?**
A. It keeps the drain from going into a state of saturation.
B. It prevents excessive current from flowing in the gate circuit.
C. It ensures that the base remains at signal ground.
D. It keeps the input signal from shorting to ground.
E. It ensures that the emitter-base junction remains reverse-biased.

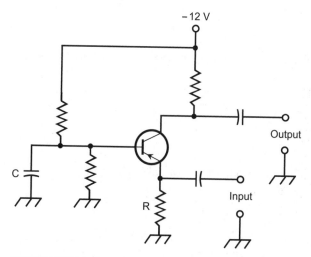

FIGURE TEST II-3 · Illustration for Part II Test Questions 16 through 18.

19. In a junction field-effect transistor (JFET), the dynamic mutual conductance specification gives us a good general idea of

   A. the range of frequencies over which the device can oscillate.
   B. how well the device can amplify signals.
   C. the optimum base bias for power amplification.
   D. the impedance transfer ratio in the source-follower configuration.
   E. the loss resistance in the base-collector junction.

20. If we want to build a tuned amplifier for operation at a frequency of 5 kHz (in the AF range), air-core inductors won't work well because they

   A. interact too much with surrounding components.
   B. exhibit too much loss.
   C. don't provide enough inductance.
   D. exhibit capacitive reactance in addition to inductive reactance.
   E. exhibit poor efficiency.

21. In a power amplifier using a PNP bipolar transistor, we define the efficiency as the

   A. ratio of the DC power input to the signal power output.
   B. ratio of the signal power output to the signal power input.
   C. sum of the signal power output and the signal power input.
   D. ratio of the signal power output to the DC power input.
   E. difference between the signal power output and the signal power input.

22. Assuming that the impedance remains constant, a signal-level gain of +4.000 dB represents an output-to-input current ratio of

   A. 1.585 to 1.
   B. 1.661 to 1.
   C. 2.000 to 1.
   D. 2.512 to 1.
   E. 4.000 to 1.

23. A signal-level gain of −4.000 dB represents an output-to-input power ratio of

   A. 1 to 1.585.
   B. 1 to 1.661.
   C. 1 to 2.000.
   D. 1 to 2.512.
   E. 1 to 4.000.

24. Figure Test II-4 is a simplified functional drawing of a

   A. bipolar transistor.
   B. JFET.
   C. MOSFET.
   D. thyrector.
   E. PIN diode.

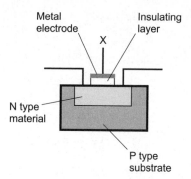

Metal
electrode

Insulating
layer

X

N type
material

P type
substrate

FIGURE TEST II-4. Illustration for
Part II Test Questions 24 through 26.

25. **In Fig. Test II-4, the terminal marked X represents the**
    A. anode.
    B. cathode.
    C. source.
    D. gate.
    E. channel.

26. **In Fig. Test II-4, the N type material forms the**
    A. base.
    B. channel.
    C. source.
    D. drain.
    E. collector.

27. **Which of the following amplifier types exhibits the *worst* linearity for the modulation envelope of an AM signal?**
    A. Class-A
    B. Class-AB$_1$
    C. Class-AB$_2$
    D. Class-B
    E. Class-C

28. **Figure Test II-5 shows a graph of the collector current versus base current for a hypothetical bipolar transistor. If we want this device to operate as a class-A amplifier, we should bias it at or near**
    A. point P.
    B. point Q.
    C. point R.
    D. point S.
    E. point T.

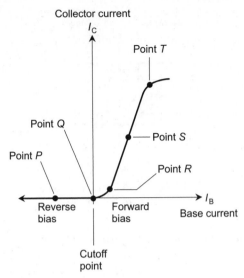

FIGURE TEST II-5 · Illustration for Part II Test
Questions 28 through 30.

29. **If we want the bipolar transistor portrayed in Fig. Test II-5 to operate as a class-B amplifier, we should bias it at or near**
   A. point P.
   B. point Q.
   C. point R.
   D. point S.
   E. point T.

30. **If we want the bipolar transistor portrayed in Fig. Test II-5 to operate as a class-C amplifier, we should bias it at or near**
   A. point P.
   B. point Q.
   C. point R.
   D. point S.
   E. point T.

31. **Suppose that we build a Hartley RF oscillator using a PNP bipolar transistor. We can adjust a variable capacitor, which we connect in parallel with an inductor, to vary the output signal's**
   A. waveform.
   B. frequency.
   C. harmonic content.
   D. amplitude.
   E. phase.

32. Suppose that we apply an input signal to the base of a grounded-emitter NPN transistor circuit, and we observe significant dynamic current gain. Then we gradually increase the signal frequency until the dynamic current gain falls to 1, meaning that the RMS output signal current equals the RMS input signal current. What's the technical term for this frequency?

    A. Alpha cutoff frequency
    B. Gamma cutoff point
    C. Saturation threshold
    D. Avalanche point
    E. Gain bandwidth product

33. What's the technical term for a linear IC that combines several different signals for transmission on a single communications channel?

    A. Mixer
    B. Differential amplifier
    C. Modem
    D. Graphic equalizer
    E. Multiplexer

34. Imagine that we built a Colpitts RF oscillator using a P-channel MOSFET and a coil whose inductance we can vary continuously over a wide range. If we get the oscillator to produce a good signal and then we adjust the coil to quadruple the inductance, we should expect that the signal frequency will

    A. remain the same.
    B. double.
    C. quadruple.
    D. become half as great.
    E. become 1/4 as great.

35. If we want to build a Clapp RF oscillator using an inductance-capacitance (LC) circuit to control the frequency, which inductor core material will result in the best frequency stability?

    A. Solid iron
    B. Air or a vacuum
    C. Laminated iron
    D. Powdered iron
    E. It doesn't matter.

36. Which of the following RF oscillator types offers the best frequency stability in most practical situations, assuming that we use components of superior quality?

    A. Pierce
    B. Colpitts
    C. Hartley
    D. Clapp
    E. It doesn't matter.

37. Which of the following terms might an engineer correctly use to describe the dynamic current gain of an NPN transistor in the grounded-emitter configuration?

A. Beta
B. Alpha
C. Gamma
D. Alpha cutoff
E. Gamma cutoff

38. Figure Test II-6 shows a circuit that uses a P-channel JFET connected to other components with the intent of building

A. an Armstrong oscillator.
B. a twin-T oscillator.
C. a Colpitts oscillator.
D. a Pierce oscillator.
E. a voltage-controlled oscillator.

39. Something is wrong with the schematic diagram of Fig. Test II-6. The oscillator won't work if we build the circuit as shown here. We can give it a better chance of generating a useful signal by

A. removing the capacitor that's in parallel with the battery.
B. short-circuiting the radio-frequency choke (RFC).
C. short-circuiting the resistor between the gate and ground.
D. replacing the P-channel JFET with a PNP bipolar transistor.
E. reversing the battery polarity.

40. The amount of feedback that occurs in the circuit of Fig. Test II-6, assuming that we modify it to ensure that it works properly, depends primarily on the

A. capacitance in parallel with the battery.
B. ratio of the capacitances across the coil.
C. inductance of the RFC.

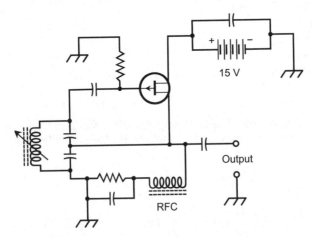

15 V

Output

RFC

FIGURE TEST II-6 · Illustration for Part II Test Questions 38 through 40.

D. value of the capacitor between the source and the output.

E. inductance of the coil.

41. **If we want to use a diode as the main element in a voltage-regulator circuit, which type of diode should we choose?**

   A. Zener

   B. Gunn

   C. PIN

   D. Point-contact

   E. Snap

42. **In a semiconductor diode, the P type wafer constitutes the**

   A. emitter.

   B. substrate.

   C. anode.

   D. cathode.

   E. collector.

43. **Figure Test II-7 is a generic diagram of a**

   A. Hartley oscillator.

   B. phase-locked loop (PLL) oscillator.

   C. Pierce oscillator.

   D. twin-T oscillator.

   E. Clapp oscillator.

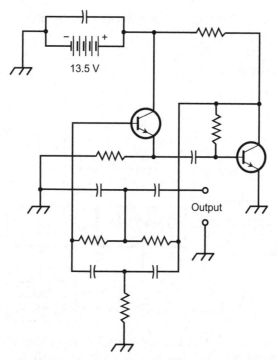

FIGURE TEST II-7 · Illustration for Part II Test Question 43.

44. **In a PNP bipolar transistor, the N type material forms the**

    A. base.

    B. gate.

    C. emitter.

    D. drain.

    E. source.

45. **Imagine that we apply two signals, one having a frequency of 7.7 MHz and the other having a frequency of 10.1 MHz, to the input of a signal mixer. What output signal frequency or frequencies will we get, not including any input that "leaks through"?**

    A. 2.4 MHz only

    B. 8.9 MHz only

    C. 17.8 MHz only

    D. 2.4 MHz and 17.8 MHz

    E. 2.4 MHz, 8.9 MHz, and 17.8 MHz

46. **Figure Test II-8 shows a simplified block diagram of a**

    A. frequency synthesizer.

    B. multivibrator.

    C. programmable power amplifier.

    D. tuned RF power amplifier.

    E. Reinartz oscillator.

47. **In the system of Fig. Test II-8, the box marked X represents**

    A. a tuned *LC* circuit.

    B. a rectifier.

    C. an envelope detector.

    D. an amplifier.

    E. a phase comparator.

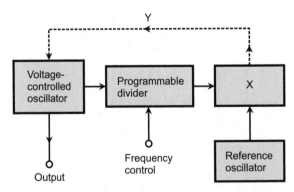

FIGURE TEST II-8 · Illustration for Part II Test Questions 46 through 48.

48. **In the system of Fig. Test II-8, the path marked Y carries**
    A. an RF signal.
    B. an error signal voltage.
    C. an AF signal.
    D. a variable reactance.
    E. nothing; it shouldn't appear in the diagram at all.

49. **What do engineers call a network of resistors designed to reduce the amplitude of a signal by a certain factor?**
    A. A damper
    B. A loss generator
    C. An attenuator
    D. An equalizer
    E. A negative gain control

50. **An optocoupler, also called an optoisolator, consists of**
    A. two LEDs or IREDs with a clear medium between.
    B. two photodiodes with a clear medium between.
    C. two photovoltaic cells with a clear medium between.
    D. an LED or IRED and a photodiode with a clear medium between.
    E. a Gunn diode and an LED or IRED with a clear medium between.

Part III

Wireless Electronics

chapter 9

Radio-Frequency Transmitters

Radio-frequency (RF) *transmitters* generate signals using oscillators, modulators, amplifiers, and other circuits with the intent of sending wireless data from one place to another in the form of an *electromagnetic* (EM) *field*.

CHAPTER OBJECTIVES

In this chapter you will

- Learn how EM fields arise, and how wavelength varies with frequency.
- Compare signal modulation modes.
- See how analog signals convert to digital signals.
- Contrast frequency modulation with phase modulation.
- Contrast fixed and full-motion image transmission methods.
- Learn how EM fields propagate in the earth's atmosphere.

The Electromagnetic Field

In an RF transmitting antenna, electrons move back and forth at extreme speed. Their velocity constantly changes as they *accelerate* in one direction, slow down, reverse direction, accelerate again, and so on.

How It Happens

When electrons move, they generate a magnetic (M) field. When electrons accelerate (speed up or slow down), they generate a *fluctuating* M field. When electrons accelerate back and forth, they generate an *alternating* M field at the same frequency as that of the AC produced by the electrons themselves.

An alternating M field gives rise to an alternating electric (E) field, which in turn spawns another alternating M field. This process repeats indefinitely, and the effect *propagates* (travels) through space at the speed of light. The E and M fields expand alternately outward from the source; at any given point in space, the lines of E flux run perpendicular to the lines of M flux. The waves propagate at right angles to both the E and M flux lines. We call this traveling phenomenon an *electromagnetic* (EM) *field*.

In the known universe, an EM field can have any conceivable frequency from centuries per cycle to quadrillions of cycles per second (or hertz). The sun has a magnetic field that oscillates with a 22-year cycle. Radio waves oscillate at thousands, millions, or billions (thousand-millions) of hertz. Infrared (IR), visible light, ultraviolet rays (UV), X rays, and gamma rays comprise EM fields that alternate at many trillions (million-millions) of hertz.

Frequency versus Wavelength

All EM fields have two important properties: the *frequency* and the *wavelength*. When we quantify them, we find that they exhibit an *inverse relation:* as one increases, the other decreases. We've already learned about AC frequency. We can express EM wavelength as a physical distance between any two adjacent points at which either the E field or the M field have identical amplitude and direction.

Let f_{MHz} represent the frequency of an EM wave in megahertz as it travels through free space (air or a vacuum). Let λ_{ft} represent the wavelength in feet. These two quantities relate according to the formula

$$\lambda_{ft} = 984/f_{MHz}$$

If we express the wavelength in meters and denote it as λ_m, then

$$\lambda_m = 300/f_{MHz}$$

The inverses of these formulas are

$$f_{\text{MHz}} = 984 / \lambda_{\text{ft}}$$

and

$$f_{\text{MHz}} = 300 / \lambda_{\text{m}}$$

In these formulas, the symbol λ is the lowercase italic Greek letter lambda, which engineers and scientists commonly use to represent wavelength.

The EM Spectrum

Physicists, astronomers, and engineers refer to the entire range of EM wavelengths as the *electromagnetic* (EM) *spectrum*. The *RF spectrum*, which includes radio, television (TV), and microwaves, constitutes a small part of the overall EM spectrum, categorized in so-called *bands* from *very low frequency* (VLF) through *extremely high frequency* (EHF) according to Table 9-1. The exact lower limit of the VLF range constitutes a subject for disagreement in the literature. Let's define it as 3 kHz, consistent with frequency boundaries between RF bands scaled by *orders of magnitude* (powers of 10).

Modulation Basics

When we *modulate* a signal, we "write" data onto an electric current or EM wave. We can carry out the *modulation* process by varying the amplitude, the frequency, or the phase of the current or wave. We can also obtain a modulated signal by

TABLE 9-1. Bands in the RF spectrum.

Frequency Designation	Frequency Range	Wavelength Range
Very Low (VLF)	3 kHz–30 kHz	100 km–10 km
Low (LF)	30 kHz–300 kHz	10 km–1 km
Medium (MF)	300 kHz–3 MHz	1 km–100 m
High (HF)	3 MHz–30 MHz	100 m–10 m
Very High (VHF)	30 MHz–300 MHz	10 m–1 m
Ultra High (UHF)	300 MHz–3 GHz	1 m–100 mm
Super High (SHF)	3 GHz–30 GHz	100 mm–10 mm
Extremely High (EHF)	30 GHz–300 GHz	10 mm–1 mm

generating a series of current or wave pulses and varying their duration, amplitude, or timing.

The Carrier

The heart of a wireless signal comprises a sine wave called the *carrier*. It can range in frequency from a few kilohertz (kHz) to many gigahertz (GHz). If we want a modulated carrier to provide us with effective, reliable data transfer, the carrier must have a frequency of at least 10 times the highest frequency of the modulating signal. We can state this principle another way: The highest frequency component in the modulating information should never exceed 10% of the carrier frequency.

On/Off Keying

The simplest form of modulation involves *on/off keying* of the carrier. We can do it in the oscillator of a radio transmitter to send *Morse code*, one of the simplest known *binary* (two-state) modulation modes, as shown in Fig. 9-1.

The duration of a Morse-code *dot* equals the duration of one *binary digit*, more often called a *bit*. A bit represents the smallest or shortest possible unit of data in a system whose only two possible conditions are fully on (the *high* state, also called *mark*) and fully off (the *low* state, also called *space*). A *dash* measures three bits long. We can build up messages of any desired length as follows:

- One space bit exists between the dots and dashes in a *character*
- Three contiguous space bits exist between the characters in a *word*
- Seven contiguous space bits exist between the words in a *sentence*
- Seven contiguous space bits follow the *period* at the end of a sentence

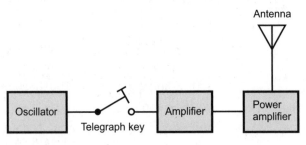

FIGURE 9-1 · Block diagram of a Morse-code transmitter that uses on/off keying.

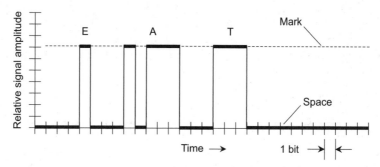

FIGURE 9-2 · The Morse-code word "EAT" as sent using on/off keying.

Figure 9-2 is a graphical amplitude-versus-time rendition of a simple three-letter word in Morse code, as it might appear on an oscilloscope when we key a source of DC.

TIP *Morse code is one of the slowest known methods of data transmission. Human operators use speeds ranging from about 5 words per minute (WPM) to 40 or 50 WPM. Machines, such as computers and data terminals, function at thousands, millions, or billions of words per minute.*

Frequency-Shift Keying

If we use *frequency-shift keying* (FSK), we can send binary digital data over wireless links at higher speeds than on/off keying allows. In some FSK systems, the carrier frequency shifts between the mark and space conditions, usually by a few hundred hertz or less. In other systems, a two-tone audio-frequency (AF) sine wave modulates the carrier, a mode called *audio-frequency-shift keying* (AFSK). The two most common codes used with FSK and AFSK are *Baudot* (pronounced "baw-DOE") and *ASCII* (pronounced "ASK-ee"). The acronym ASCII stands for *American Standard Code for Information Interchange*.

In *radioteletype* (RTTY) FSK and AFSK systems, a *terminal unit* (TU) converts the digital signals into electrical impulses that display characters on a computer screen or teleprinter. The TU also generates the signals necessary to send RTTY when an operator types on a keyboard. A specialized TU that sends and receives AFSK is called a *modem*, an acronym that stands for *modulator/demodulator*.

With FSK or AFSK, we get fewer errors than we do with simple on/off keying because the spaces are positively identified, rather than existing as mere gaps in the data. A sudden noise burst in an on/off keyed signal can confuse a

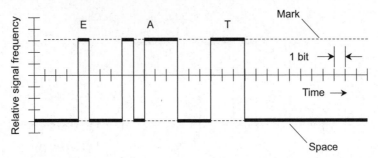

FIGURE 9-3 · The Morse-code word "EAT" as sent using FSK.

receiver into reading a space as a mark, but when the space has its own definitive signal, this type of error happens less often. Figure 9-3 shows a frequency-versus-time rendition of a simple three-letter word transmitted using Morse code in FSK mode. Figure 9-4 is a block diagram of an AFSK transmitter.

Amplitude Modulation

A voice signal presents itself as a complex waveform with frequencies mostly in the range between 300 Hz and 3000 Hz. We can vary, or modulate, the instantaneous amplitude of a carrier with a voice waveform. Figure 9-5 shows a bipolar-transistor circuit for obtaining *amplitude modulation* (AM) that works well as long as the audio input signal doesn't get too strong. If the audio input signal exceeds a certain amplitude, distortion occurs.

Figure 9-6 shows block diagrams of two different AM wireless transmitters. Figure 9-6A portrays an example of *low-level AM*, in which all amplifiers

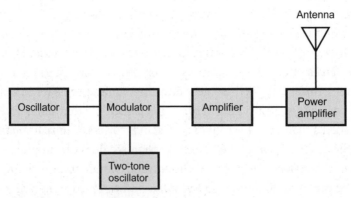

FIGURE 9-4 · Block diagram of an AFSK transmitter.

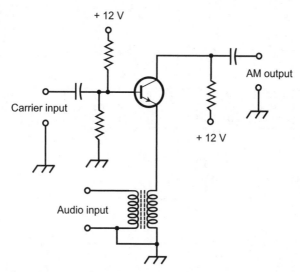

FIGURE 9-5 · A simple circuit for generating an amplitude-modulated (AM) signal.

following the modulator have excellent linearity so that they don't distort the modulating waveform. In *high-level AM*, the modulation takes place in the last power amplifier (PA) stage as shown in Fig. 9-6B. The PA operates in class C for maximum efficiency, and serves as both the modulator and the final amplifier. Because no amplification stages exist beyond the modulator, we don't have to worry about linearity in the amplifiers.

Modulation Percentage and Sidebands

We can express the extent of AM as a percentage from 0% (an unmodulated carrier) to 100% (the maximum modulation we can get without distortion). In an AM signal modulated 100%, only 1/3 of the signal power conveys useful data. The carrier wave consumes the remaining 2/3 of the signal power. Increasing the modulation past 100% causes distortion, degrades the transmitter's efficiency, and causes the signal to "spread itself out" over an unnecessarily wide span (or band) of frequencies, a phenomenon called *splatter*.

Figure 9-7 shows a *spectral display* of an AM voice radio signal. The horizontal scale is calibrated in increments of 1 kHz per division. The vertical scale is calibrated in decibels relative to 1 mW of signal power (dBm), with 0 dBm at the top. The AF components, containing the intelligence transmitted, appear as *sidebands* on either side of the carrier. The RF energy between –3 kHz and the carrier frequency constitutes the *lower sideband*, abbreviated LSB. The RF

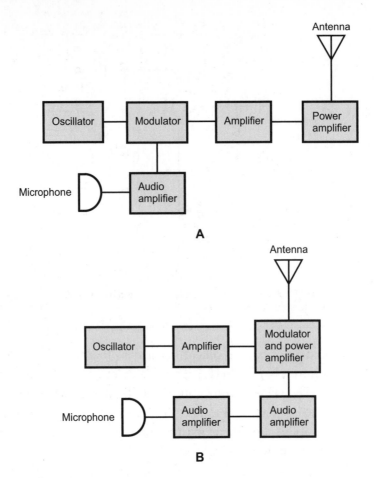

FIGURE 9-6 · At A, low-level AM. At B, high-level AM.

energy between the carrier frequency and +3 kHz represents the *upper side-band*, abbreviated USB. (Don't confuse this use of the abbreviation USB with the *universal serial bus* translation in personal computing!) The *bandwidth* of the RF signal equals the difference between the maximum and minimum side-band frequencies.

In an AM signal, the bandwidth always equals twice the highest audio modulating frequency. In the example of Fig. 9-7, the voice energy remains at or below 3 kHz, so the bandwidth of the complete signal equals 6 kHz, typical of an analog voice communications signal. In AM broadcasting that includes music, images, or high-speed data, the energy occupies bandwidth somewhat greater than 6 kHz.

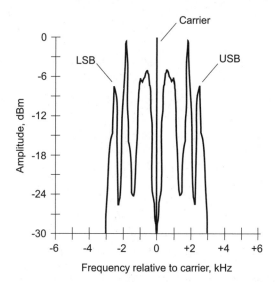

FIGURE 9-7 · Spectral display of a typical AM voice signal.

Single Sideband

In AM, the carrier wave consumes at least 2/3 of the RF signal power, and the sidebands exist as mirror-image duplicates. Therefore, AM is redundant and inefficient. If we could get rid of the carrier and one of the sidebands, we'd convey all the same information while consuming less power. Alternatively, we could get a stronger signal for a given amount of RF power, and reduce the signal bandwidth to a little less than half that of an AM signal modulated with the same data. The resulting *spectrum savings* would allow us to fit more than twice as many signals into a specific band of frequencies. During the early and middle parts of the 20th century, communications engineers perfected a way to modify AM signals in this way. They called the resulting mode *single sideband* (SSB), a term that has endured to the present day.

When we remove the carrier and one of the sidebands from an AM signal, the remaining RF energy has a spectral display resembling Fig. 9-8. In this case, we have eliminated the USB energy along with the carrier from the AM signal shown in Fig. 9-7, leaving energy only in the LSB. We could, however, just as well remove the LSB energy along with the carrier, leaving energy only in the USB.

We can suppress the carrier in an AM signal with a *balanced modulator*, a circuit that has two transistors with the inputs connected in push-pull and the

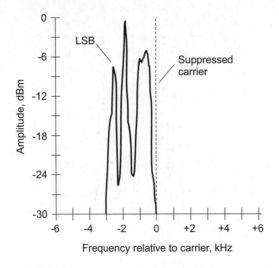

FIGURE 9-8 · Spectral display of a typical LSB voice signal.

outputs connected in parallel as shown in Fig. 9-9. This arrangement "cancels out" the carrier wave in the output signals, leaving only LSB and USB energy yielding a *double-sideband suppressed-carrier* (DSBSC) signal, often called simply *double sideband* (DSB). Immediately following the balanced modulator, a

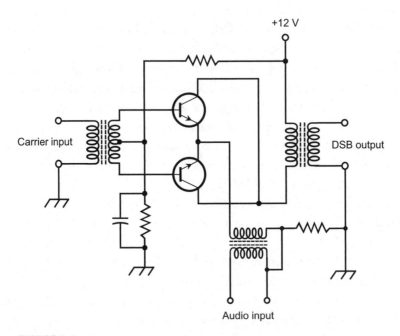

FIGURE 9-9 · A balanced modulator circuit using bipolar transistors.

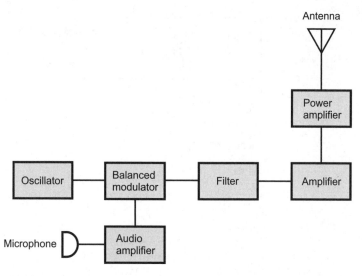

FIGURE 9-10 · Block diagram of an SSB transmitter.

bandpass filter suppresses one of the sidebands by "brute force" to obtain an SSB signal. This type of filter allows signals to get through easily inside a *band* of frequencies between a certain minimum and maximum, but attenuates signals outside that band.

Figure 9-10 is a block diagram of a simple SSB transmitter. The balanced modulator operates in a low-power section. The RF amplifiers that follow any type of amplitude modulator, including a balanced modulator, must all operate in a linear manner to prevent distortion and splatter. The RF amplifiers in an SSB transmitter generally work in class A except for the final PA, which operates in class AB_1, class AB_2, or class B. We'll never see a class-C amplifier serving as the PA in an SSB transmitter because that mode introduces nonlinearity into the modulation envelope of any signal whose amplitude varies over a continuous range.

Frequency and Phase Modulation

In *frequency modulation* (FM), the instantaneous amplitude of a signal remains constant; the instantaneous frequency varies instead. A nonlinear PA such as a class-C amplifier can function in an FM transmitter without causing signal distortion because the amplitude always stays the same.

We can obtain FM by applying an audio signal to the varactor diode in a voltage-controlled oscillator (VCO). Figure 9-11 shows an example of this scheme, known as *reactance modulation*, for use with a Colpitts oscillator

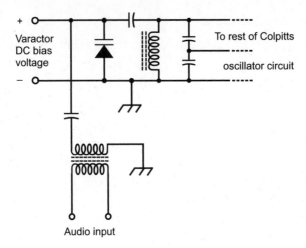

FIGURE 9-11 · Reactance modulation to obtain FM.

circuit. The varying voltage across the varactor changes its capacitance "in sync" with the audio waveform. The fluctuating capacitance causes variations of the resonant frequency of the inductance-capacitance (*LC*) tuned circuit, causing small fluctuations in the frequency generated by the oscillator.

We can obtain FM indirectly by modulating the phase of an oscillator signal. When we vary the phase from instant to instant, we provoke small instantaneous variations in the frequency. In this mode, called *phase modulation* (PM), we must process the audio signal before we apply it to the modulator, adjusting the *frequency response* of the audio amplifiers. Otherwise the signal will sound distorted when we listen to it in a receiver designed for ordinary FM.

Still Struggling

You can't change an object's location without also changing its speed, and you can't change an object's speed without changing its location! Similarly, you can't alter phase without altering frequency (or vice-versa). Nevertheless, frequency modulation and phase modulation differ from each other in a fundamental sense, just as position and motion do.

Deviation and Modulation Index

In an FM or PM signal, the *deviation* quantifies the maximum extent to which the instantaneous carrier frequency differs from the unmodulated-carrier frequency. In most FM and PM voice-communications transmitters, the deviation equals ±5.0 kHz when we apply a sine-wave audio tone at a frequency of 1 kHz (Fig. 9-12). We call this mode *narrowband FM* (NBFM). The bandwidth of an NBFM signal roughly equals that of an AM signal containing the same modulating information. In FM music broadcasting, image transmission, and high-speed data applications, the deviation exceeds ±5.0 kHz, giving us *wideband FM* (WBFM).

For any particular oscillator frequency, the maximum deviation obtainable with direct FM exceeds the maximum deviation that we can get with PM. However, we can increase the deviation of any FM or PM signal with a *frequency multiplier*. When an FM or PM signal passes through a frequency multiplier, the deviation increases by the same factor as the carrier frequency increases.

The deviation in an FM or PM signal should *at least* equal the highest modulating audio frequency if we want optimum audio fidelity. Therefore, ±5.0 kHz represents more than enough deviation for voice communications, which only needs audio frequencies up to 3 kHz. For music, we need deviation of ±15 kHz or ±20 kHz for good sound reproduction in a receiver. (The human hearing range extends up to 15 or 20 kHz.)

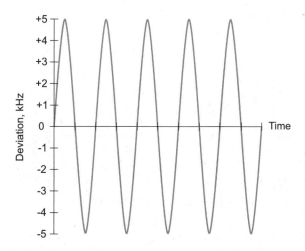

FIGURE 9-12 · Frequency-versus-time rendition of a narrowband FM signal, typical of voice transmissions.

In any FM signal, engineers call the ratio of the frequency deviation to the highest modulating audio frequency the *modulation index*. Ideally, this figure lies between 1:1 and 2:1. If it's less than 1:1, the signal sounds muffled or distorted, and efficiency is sacrificed. Increasing the modulation index much beyond 2:1 broadens the bandwidth without providing significant improvement in intelligibility or fidelity.

Pulse Modulation

We can modulate a communications signal by varying some aspect of a constant stream of signal pulses. Figure 9-13 shows several types of *pulse modulation* as amplitude-versus-time graphs. The modulating waveform in each case appears as a dashed curve, and the pulses appear as vertical gray bars.

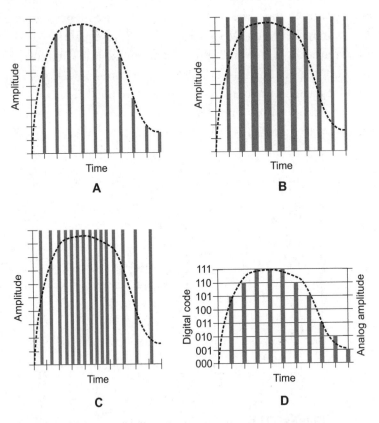

FIGURE 9-13 · At A, pulse amplitude modulation; at B, pulse width modulation; at C, pulse interval modulation; at D, pulse code modulation. Dashed curves represent the analog input. Vertical bars represent output pulses.

Pulse-Amplitude Modulation

In *pulse-amplitude modulation* (PAM), the strength of each individual pulse varies according to the instantaneous amplitude of the modulating waveform. In this respect, PAM resembles AM. Figure 9-13A shows a hypothetical PAM signal. Normally, the pulse amplitude increases as the instantaneous modulating-signal level increases (*positive PAM*). But we can reverse this relation so that higher audio levels produce lower pulse amplitudes (*negative PAM*). In that case, the signal pulses are strongest when no audio input exists. The RF stages of the transmitter must "work harder" to produce negative PAM than to generate positive PAM.

Pulse-Width Modulation

We can modulate the output of an RF transmitter by varying the width (duration) of the pulses to obtain *pulse-width modulation* (PWM), also known as *pulse duration modulation* (PDM), as shown in Fig. 9-13B. Normally, the pulse width increases as the instantaneous modulating-signal level increases (*positive PWM*). But this can be reversed (negative *PWM*). The transmitter must "work harder" to accomplish negative PWM than to generate positive PWM. In either case, the pulse amplitude remains constant.

Pulse-Interval Modulation

Even if all the pulses have the same amplitude and the same duration, we can obtain pulse modulation by varying how often the pulses occur. In PAM and PWM, we always transmit the pulses at the same time interval, known as the *sampling interval*. But in *pulse interval-modulation* (PIM), pulses can occur more or less frequently than they do under conditions of no modulation. Figure 9-13C shows a hypothetical PIM signal. Every pulse has the same amplitude and the same duration, but the time interval between them changes. When there is no modulating signal, the pulses emerge from the transmitter evenly spaced with respect to time. An increase in the instantaneous data amplitude might cause pulses to be sent more often, as is the case in Fig. 9-13C (*positive PIM*). Alternatively, an increase in instantaneous data level might slow down the rate at which the pulses emerge (*negative PIM*).

Pulse-Code Modulation

Nowadays, most data transmitters use digital modes in which the modulating data attains only certain defined states. Digital transmission offers better efficiency than analog transmission, in which the modulating data varies continuously (in theory attaining infinitely many different states). Digital communications allows for improved *signal-to-noise* (S/N) *ratio*, narrower signal bandwidth, better accuracy, and superior reliability as compared to analog modes that convey the same information at the same speed.

In *pulse-code modulation* (PCM), any of the above-described aspects—amplitude, width, or interval—of a pulse sequence (or *pulse train*) can be varied. But rather than having infinitely many possible states, the signal can attain only a few states. The number of states generally equals some power of 2, such as 2^2 (4 states), 2^3 (8 states), 2^4 (16 states), 2^5 (32 states), 2^6 (64 states), and so on. As we increase the number of possible signal levels, the overall fidelity and data transmission speed improve, but the signal gets more complicated and takes up more bandwidth. Figure 9-13D shows an example of 8-level PCM.

Analog-to-Digital Conversion

Pulse-code modulation, such as we see graphically at Fig. 9-13D, constitutes a common form of *analog-to-digital* (A/D) *conversion*. A voice signal, or any continuously variable signal, can be *digitized*, or converted into a train of pulses whose amplitudes can achieve only certain defined levels.

In A/D conversion, because the number of digital states always equals some power of 2, we can represent the output signal as a binary-number code. Fidelity improves as the exponent increases. Engineers call the number of digital states the *sampling resolution*, or simply the *resolution*. A resolution of $2^3 = 8$ (as shown in Fig. 9-13D) works quite well for voice communications. A resolution of $2^4 = 16$ can provide fairly decent music reproduction.

The efficiency with which we can digitize a signal depends on the frequency at which we carry out the sampling. In general, the *sampling rate* must equal at least twice the highest data frequency. For an audio signal with components as high as 3 kHz, the minimum sampling rate for effective digitization is 6 kHz. For music it's higher; for images, it's higher still.

PROBLEM 9-1

Suppose that we generate an FM signal indirectly with phase modulation to obtain an effective frequency deviation of ±2.5 kHz. How can we increase the deviation to the standard ±5.0 kHz for voice transmission?

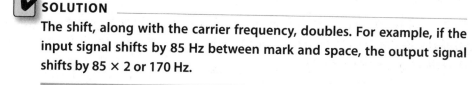

SOLUTION

We can send the signal through a *frequency doubler*, a circuit that produces an output signal having twice the frequency of the input signal. This circuit doubles the deviation along with the signal frequency.

PROBLEM 9-2

What happens to an FSK signal when it passes through a frequency doubler?

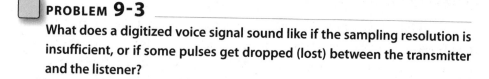

SOLUTION

The shift, along with the carrier frequency, doubles. For example, if the input signal shifts by 85 Hz between mark and space, the output signal shifts by 85 × 2 or 170 Hz.

PROBLEM 9-3

What does a digitized voice signal sound like if the sampling resolution is insufficient, or if some pulses get dropped (lost) between the transmitter and the listener?

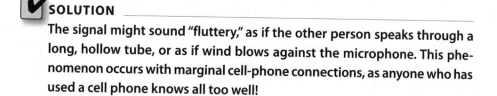

SOLUTION

The signal might sound "fluttery," as if the other person speaks through a long, hollow tube, or as if wind blows against the microphone. This phenomenon occurs with marginal cell-phone connections, as anyone who has used a cell phone knows all too well!

Image Transmission

Nonmoving images can be sent within the same bandwidth as voice signals. For high-resolution, moving images such as video files, we need more bandwidth than we do for low-resolution, nonmoving images such as simple drawings.

Facsimile

We can transmit nonmoving images (also called *still images*) by *facsimile*, also called *fax*. If we send the data slowly enough, we can render high detail within a 3-kHz-wide band, the standard for voice communications. This flexibility explains why detailed fax images can effectively propagate over a *plain old telephone service* (POTS) line.

In an electromechanical fax machine, a paper document or photo is wrapped around a *drum*. The drum rotates at a slow, controlled rate. A spot of light scans in narrow horizontal lines across the document. The drum moves the document so that a single line scans with each pass of the light spot. This process continues, line by line, until the device has scanned the complete *frame* (image). A *photodetector* picks up the light rays reflected from the document paper. Dark portions of the image reflect less light than bright parts, so the current through the photodetector varies as the light beam passes over various regions. This current modulates a carrier in one of the modes described earlier, such as AM, FM, or SSB.

TIP   *A fax receiver duplicates the transmitter's scanning rate and pattern, and a display or printer reproduces the image in grayscale (shades of gray ranging from black to white, without color).*

Slow-Scan Television

We can think of *slow-scan television* (SSTV) as a fast, repetitive form of fax. An SSTV signal, like a fax signal, propagates within a band of frequencies as narrow as that a human voice. And, like fax, SSTV transmission reproduces only still images, not moving ones. However, an SSTV system scans and transmits a full image in 8 seconds, rather than the minute or more typical of fax. With SSTV, we can offer our viewers some sense of motion in a scene. However, the speed bonus comes with a tradeoff: In SSTV, we get lower *resolution*, meaning less image detail, than we can obtain with fax.

An SSTV frame contains 120 lines. The darkest parts of the image are sent at 1.5 kHz and the brightest parts are sent at 2.3 kHz. *Synchronization (sync) pulses*, which keep the receiving apparatus in step with the transmitter, are sent at 1.2 kHz. A *vertical sync pulse* tells the receiver that it's time to begin a new frame; this pulse lasts for 30 milliseconds (ms). A *horizontal sync pulse* tells the receiver to start a new line in a frame; its duration equals 5 ms. These pulses prevent *rolling* (haphazard vertical image motion) or *tearing* (lack of horizontal synchronization).

TIP   *We can program any personal computer so that its monitor functions as an SSTV display. With some effort, we can also find converters that allow us to look at SSTV signals on a conventional TV set.*

Fast-Scan Television

Old-fashioned analog television is also known as *fast-scan TV* (FSTV). The FSTV broadcast scheme was originally developed by the National Television

Systems Committee (NTSC) in 1953. This technology, also called NTSC TV, was regarded as a breakthrough at that time. While broadcasters no longer use FSTV, some amateur ("ham") radio operators still do. The frames are transmitted at the rate of 30 per second. A single, complete frame contains 525 lines. The short frame duration and increased resolution make it necessary to use a much wider frequency band than fax or SSTV modes require. A typical video FSTV signal takes up 6 MHz of spectrum space.

Fast-scan TV signals can be transmitted using conventional AM or wideband FM. With AM, one of the sidebands can be almost completely eliminated using a bandpass filter, leaving only the carrier and the other sideband fully intact. Engineers call this mode *vestigial sideband* (VSB) transmission. It cuts the bandwidth of an FSTV signal down to a little more than 3 MHz.

Because of the large amount of spectrum space needed to send FSTV, this mode isn't practical for use at frequencies below about 30 MHz. In the "olden days of TV," all commercial FSTV transmission was done above 50 MHz, with the great majority of channels having frequencies far higher than that. Channels 2 through 13 on an old-fashioned FSTV receiver are sometimes called the *very high frequency* (VHF) *channels*; the higher channels are called the *ultra high frequency* (UHF) *channels*.

Figure 9-14 portrays a time-domain graph of the waveform of a single line in an FSTV video signal. This graph shows us 1/525 of a complete frame. The

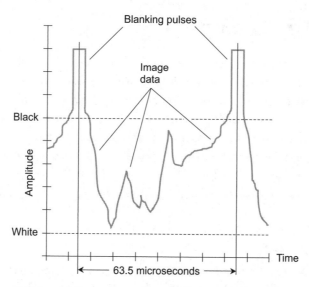

FIGURE 9-14 · A conventional analog FSTV signal, showing one line of data.

highest instantaneous signal amplitude corresponds to the darkest shade, and the lowest amplitude corresponds to the lightest shade. Therefore, the FSTV signal is sent "negatively." This convention allows *retracing* (moving from the end of one line to the beginning of the next) to synchronize between the transmitter and receiver. A well-defined, strong *blanking pulse* tells the receiver when to retrace, and it also shuts off or "blacks out" the beam during the brief moment while the receiver display retraces.

Color FSTV works by sending three separate monochromatic signals, corresponding to the *primary colors* red, blue, and green. The signals are, in effect, black-and-red, black-and-blue, and black-and-green. The receiver recombines these signals and displays the resulting video as a matrix of red, blue, and green dots. When viewed from a distance, the dots are too small to be individually discernible. Various combinations of red, blue, and green intensities result in reproduction of all possible hues and saturations of color.

An FSTV transmitter consists of a camera, an oscillator, an amplitude modulator, and a series of amplifiers for the video signal. The audio system consists of an input device (such as a microphone), an oscillator, a frequency modulator, and a feed system that couples the RF output into the video amplifier chain. The complete system also has an antenna or cable output. Figure 9-15 shows a block diagram of an analog FSTV transmitter.

Still Struggling

If you watched much commercial analog TV "in the olden days" using indoor or roof-mounted antennas, did you notice that weak TV signals had poor *contrast*, looking "washed-out"? Faint blanking pulses resulted in incomplete retrace blanking, making the blackest parts of the image appear gray instead. But this little problem was better than having the TV receiver lose track of the retrace signals, as might have happened if the highest instantaneous amplitude went with the lightest image shade.

PROBLEM 9-4

How can we cut the bandwidth of a conventional analog FSTV video signal roughly in half without sacrificing any of the image detail, and without slowing down the scanning speed?

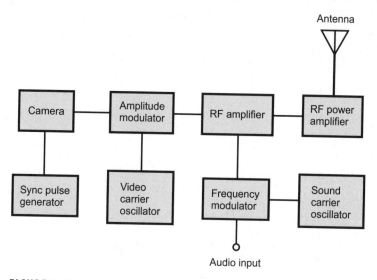

FIGURE 9-15 · Block diagram of a conventional analog FSTV transmitter.

✓ SOLUTION

A conventional analog FSTV video transmitter uses AM, so its output contains a carrier wave along with the LSB and USB. If we convert the signal to SSB, its bandwidth would, in theory, go down from 6 MHz to a little less than 3 MHz, slightly narrower than the bandwidth of a vestigial-sideband signal.

PROBLEM 9-5

Suppose that we point an FSTV camera at a featureless, gray, overcast sky or a blank gray wall. What does a time-domain display of the video signal, such as we would see on the screen of an oscilloscope, look like in this situation?

✓ SOLUTION

The video signal amplitude between blanking pulses remains constant, somewhere between white and black. When we analyze the signal with an oscilloscope, we see something like Fig. 9-16.

High-Definition Television

The term *high-definition television* (HDTV) refers to any of several methods for getting more detail into a TV picture than could ever be done with FSTV.

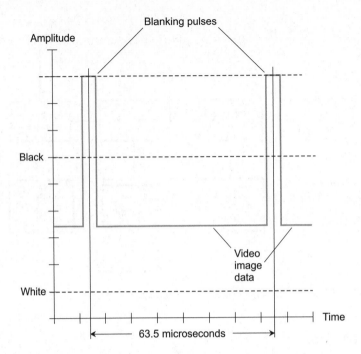

FIGURE 9-16 · Illustration for Problem 9-5.

The HDTV mode also offers superior sound quality for a more satisfying home TV and home theater experience.

The main challenges for the developers and marketers of HDTV in its early years were getting the general public to want it, and offering it to people at an affordable price. Since the turn of the millennium, a large segment of the market has embraced HDTV. The overall quality of conventional TV receivers has gotten better as well, and the transition from analog to digital broadcasting has accelerated this improvement.

An analog FSTV picture (now obsolete) had 525 lines per frame, but HDTV systems have many more. The HDTV image is scanned about 60 times per second. High-definition TV is sent in a digital mode, offering another advantage over old-fashioned analog FSTV. Digital signals propagate better, are easier to deal with when weak, and can be processed using technologies that analog signals defy.

Some HDTV systems use a technique called *interlacing* to double the number of lines in each frame, without the need for extremely sophisticated scanning apparatus. An interlaced system scans the screen twice for each complete video image, so there are actually two overlapping images instead of one image. The two images, called *rasters*, are "meshed" together, effectively doubling the image resolution without doubling the cost of manufacture. Unfortunately, this process can cause annoying "jitter" in the reproduction of fast-moving scenes or special effects.

Radio-Wave Propagation

The fact that RF waves can travel long distances, even through a vacuum and in the absence of any supporting infrastructure, has fascinated scientists ever since Marconi and Tesla discovered the phenomenon around the year 1900. Let's look at a few of the factors that affect wireless communications at radio frequencies.

Polarization

We define the *polarization* of an EM wave as the orientation of the electric-field (E-field) lines of flux. If the E-field flux lines run parallel to the earth's surface, we have *horizontal polarization*. If the E-field flux lines run perpendicular to the earth's surface, we have *vertical polarization*. Polarization can also have a "slant," of course.

In some situations, the E-flux lines rotate as the wave travels through space. In a situation of that sort, we have *circular polarization* if the E-field intensity remains constant and the orientation rotates at a constant angular speed. If the E-field intensity is more intense in some planes than in others, or if the angular rotation rate is greater in some planes than in others, we have *elliptical polarization*.

A circularly or elliptically polarized wave can appear to turn either *clockwise* or *counterclockwise* as the wavefronts approach us. The rotational direction of an approaching wave is called the *sense* of polarization. Some engineers use the term *right-hand* instead of clockwise and the term *left-hand* instead of counterclockwise.

The Line-of-Sight Wave

Electromagnetic waves follow straight lines unless something bends them. *Line-of-sight* propagation can take place even when we can't literally see the receiving antenna from the transmitting antenna's location, because radio waves penetrate nonconducting objects, such as trees and frame houses, to some extent. The line-of-sight wave consists of two components, called the *direct wave* and the *reflected wave*.

In a direct wave, the longest wavelengths are least affected by obstructions. At very low, low, and medium frequencies, direct waves can *diffract* around physical objects, even if those objects are *radio-opaque*. As the frequency rises, especially above about 3 MHz, obstructions have a greater and greater blocking effect.

Electromagnetic waves reflect from the earth's surface and from conducting objects, such as wires, broadcast towers, water towers, and steel beams. The reflected wave always travels farther than the direct wave. The two waves rarely arrive at the receiving antenna in exact phase coincidence.

If the direct and reflected waves arrive at the receiving antenna with equal strength but 180° out of phase, a *dead spot* occurs. This phenomenon is most noticeable at the highest frequencies. At VHF and UHF, an improvement in reception can sometimes result from moving the transmitting or receiving antenna only a few centimeters!

TIP   *In mobile operation, when the transmitter and/or receiver constantly move, dead spots produce rapid, repeated interruptions in the received signal, a phenomenon that "ham" radio operators know quite well; they call it* **picket fencing.**

The Surface Wave

At AC frequencies below about 10 MHz, the earth's surface constitutes a fairly good conductor. Because of this, vertically polarized EM waves follow the surface for hundreds or even thousands of kilometers, with the earth helping to transmit the signals. As we reduce the frequency, we observe decreasing *ground loss*, and the waves can travel progressively greater distances by means of *surface-wave propagation*. Horizontally polarized waves don't travel well in this mode because the conductive surface of the earth effectively short-circuits horizontal E-field flux. At frequencies above about 10 MHz, the earth becomes lossy, and surface-wave propagation rarely occurs over distances greater than a few kilometers.

Sky-Wave EM Propagation

Ionization in the upper atmosphere, caused by solar radiation, can return EM waves to the earth at certain frequencies. The ionization takes place in four distinct layers, as follows.

1. The D layer exists at an altitude of about 50 km (roughly 30 mi), and ordinarily exists only on the daylight side of the planet. It absorbs radio waves at some frequencies, impeding long-distance ionospheric propagation.

2. The *E layer*, which lies about 80 km (roughly 50 mi) above the surface, also exists mainly during the day, although nighttime ionization is sometimes observed. The E layer can provide medium-range radio communication at certain frequencies.

3. The F_1 layer, normally present only on the daylight side of the earth, forms at about 200 km or 125 mi altitude. It can provide long-distance communication.

4. The F_2 layer lies roughly 300 km or 180 mi above most, or all, of the earth. It, like the F_1 layer, facilitates long-distance communication.

Communication by F_1- and F_2-layer propagation is nearly always possible between any two points on the earth within a certain range of frequencies between 5 MHz and 30 MHz. This range constantly varies depending on atmospheric conditions, solar activity, the time of day, and the locations of the transmitting and receiving stations. The lowest frequency in this range is the *minimum usable frequency*, while the highest frequency in the range constitutes the *maximum usable frequency*.

TIP   *Some engineers and amateur radio operators ignore the distinction between the F_1 and F_2 layers, speaking of them together as the* **F** *layer.*

Tropospheric Propagation

At frequencies above about 30 MHz, the lower atmosphere bends radio waves toward the surface. *Tropospheric bending* occurs because the *index of refraction* of air, with respect to EM waves, decreases as the altitude increases. This effect makes it possible for two wireless stations to maintain contact over spans of many kilometers, even when the ionosphere will not return waves to the earth.

Ducting constitutes another type of tropospheric propagation, which happens less often than bending but produces more dramatic effects. Ducting takes place when EM waves become "trapped" in a layer of cool, dense air sandwiched between two layers of warmer air. Ducting, like bending, occurs almost entirely at frequencies above 30 MHz.

Still another tropospheric-propagation mode is known as *troposcatter*. This phenomenon takes place because air molecules, dust grains, and water droplets scatter some of the EM field, in much the same way as smoke scatters rays of light. We observe troposcatter at the same frequencies as we witness bending and ducting. Troposcatter always occurs to some extent, regardless of weather conditions—even in clear air.

TIP   *Tropospheric propagation in general, without mention of the specific mode, is sometimes called* **tropo.**

Auroral Propagation

In the presence of unusual solar activity, the *aurora borealis* ("northern lights") or *aurora australis* ("southern lights") can return radio waves to the earth, allowing so-called *auroral propagation*. The aurora occur at altitudes of about 65 to 400 km (roughly 40 to 250 mi) in circular zones surrounding the earth's magnetic poles. Auroral propagation can occur between any two points on the earth's surface from which the same part of the aurora lies on a line of sight. Auroral propagation seldom takes place when either the transmitting station or the receiving station is located in the tropics, but people who live in the Arctic, or in extreme southern South America and Australia, commonly witness it.

Auroral propagation causes rapid and deep signal fading, which nearly always renders analog voice and video signals unintelligible. Digital modes are somewhat more effective for communication by means of auroral propagation, but the carrier is often spread out over several hundred hertz as a result of random phase modulation induced by auroral motion. This "spectral smearing effect" severely limits the maximum data transfer rate. Auroral propagation commonly accompanies poor ionospheric propagation resulting from sudden eruptions on the sun's surface called *solar flares*.

Meteor-Scatter Propagation

Meteors produce *ionized trails* that persist for lengths of time ranging from a fraction of a second up to several seconds. The exact duration of a particular trail depends on the size of the responsible meteor, its speed, and the angle at which it enters the atmosphere. A single meteor trail rarely lasts long enough to allow transmission of much data. However, during a *meteor shower*, multiple trails can produce almost continuous ionization for a period of hours. So many trails occur, at such frequent intervals, that a sort of "artificial ionospheric layer" results. Ionized regions of this type can reflect radio waves at certain frequencies. Communications engineers call this effect *meteor-scatter propagation*, or sometimes simply *meteor scatter*. It can take place at frequencies far above 30 MHz and over distances up to about 2400 km or 1500 mi.

TIP *In meteor-scatter mode, the maximum obtainable communications range depends on the altitude of the ionized trail, and also on the relative spatial locations of the trail, the transmitting station, and the receiving station.*

Moonbounce

The moon orbits the earth at a distance of about 400,000 km (250,000 mi), close enough to return signals from transmitters affordable to private citizens.

Amateur radio operators have successfully carried out *earth-moon-earth* (EME) communications, also called *moonbounce*, in the VHF and UHF portions of the radio spectrum. Successful EME communications requires a sensitive receiver using a low-noise *preamplifier*, a large, directional antenna, and a high-power transmitter. Digital modes work better than analog modes.

Signal *path loss* presents the most outstanding difficulty for anyone who contemplates EME communications. The received signals are always weak. High-gain directional antennas must constantly stay aimed at the moon, a requirement that dictates the use of steerable antenna arrays, preferably with rotators governed by computers programmed to follow the path of the moon across the sky. The EME *path loss* increases with increasing frequency, but this effect is offset by the more manageable size of high-gain antennas as the wavelength decreases.

Moonbounce communications can be done by day as well as by night, but *solar noise* can pose a problem with daytime efforts. Difficulties peak near the time of the new moon, when the moon lies near a line between the earth and the sun. The sun is an immensely powerful, broadband generator of EM energy, and its noise can drown out faint human-made EME signals. Problems can also occur with *cosmic noise* when the moon passes near "noisy" regions in the radio sky. The constellation *Sagittarius* lies in the direction of the center of the Milky Way galaxy, and EME performance suffers when the moon passes in front of this region.

The moon keeps the same face more or less toward the earth at all times, but some back-and-forth wobbling occurs. This motion produces rapid, deep fluctuations in signal strength, a phenomenon, known as *libration fading*, that becomes increasingly pronounced as we increase the operating frequency. Libration fading results from the return of multiple signals that vary rapidly in phase with respect to each other, having bounced back to us from multiple points on the moon's surface.

Still Struggling

Take careful note of the spelling and meaning of the word *libration* (pronounced "lie-BRAY-shun"), which refers to physical wobbling motion. Don't confuse libration with *liberation*, which refers to the attainment of human rights, or *libation*, which refers to the serving or consumption of alcoholic drinks!

QUIZ

Refer to the text in this chapter if necessary. A good score is eight correct. The answers are listed in the back of the book.

1. In an FM signal, what do we call the maximum extent to which the instantaneous carrier frequency differs from the unmodulated-carrier frequency?

    A. Bandwidth
    B. Modulation index
    C. Deviation
    D. Sampling resolution

2. What's the frequency of an RF signal whose free-space wavelength equals 12.0 ft?

    A. 25.0 MHz
    B. 41.0 MHz
    C. 82.0 MHz
    D. 150 MHz

3. The circuit shown in Fig. 9-17 could serve as

    A. an amplitude converter, although it contains an error.
    B. a frequency modulator, although it contains an error.
    C. a phase modulator, although it contains an error.
    D. a balanced modulator, although it contains an error.

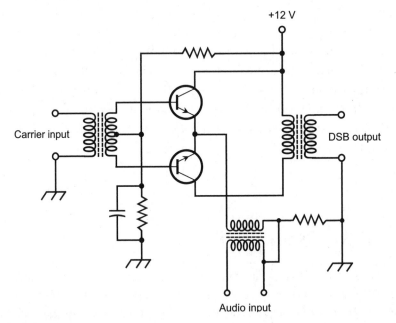

FIGURE 9-17 · Illustration for Quiz Questions 3 and 4.

4. **How can we correct the error in Fig. 9-17?**
   A. Connect the center tap in the secondary of the carrier input transformer through a capacitor to ground, rather than to the point shown in the figure.
   B. Connect the collector of the top transistor to the bottom end of the output transformer primary, rather than to the top end as shown.
   C. Change the power-supply polarity from positive to negative, and replace the NPN bipolar transistors with N-channel JFETs.
   D. Remove the resistor from between the audio input transformer secondary and ground.

5. **Which of the following radio-wave propagation modes should we expect to encounter at a frequency of 150 MHz?**
   A. F_2-layer propagation
   B. Tropospheric bending
   C. Surface-wave propagation
   D. All of the above

6. **What's the free-space wavelength of an RF signal whose frequency equals 2.00 MHz?**
   A. 150 m
   B. 234 m
   C. 492 m
   D. 600 m

7. **Figure 9-18 is a graphical rendition of**
   A. pulse-code modulation.
   B. pulse-interval modulation.
   C. pulse-width modulation.
   D. pulse-duration modulation.

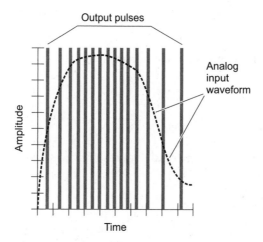

FIGURE 9-18 · Illustration for Quiz Question 7.

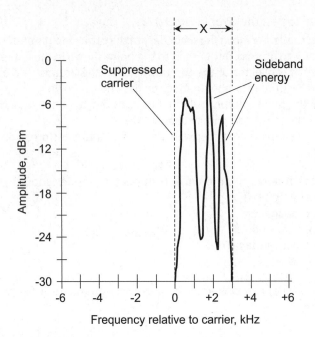

FIGURE 9-19 · Illustration for Quiz Questions 9 and 10.

8. **Suppose that we build a phase-modulator circuit that gives us an effective deviation of ±1250 Hz. We want to increase the deviation to the standard ±5.0 kHz for voice FM transmission. How can we do it?**

   A. We can't.

   B. We can put the signal through a programmable divider set to a factor of 4.

   C. We can mix the signal with another signal having 4 times the frequency.

   D. We can put the signal through a nonlinear amplifier with an output *LC* circuit tuned to 4 times the input frequency.

9. **Figure 9-19 is a frequency-domain portrayal of**

   A. a USB signal.

   B. an LSB signal.

   C. a DSB signal.

   D. an AM signal.

10. **In Fig. 9-19, the parameter marked X represents the signal's**

   A. deviation.

   B. modulation index.

   C. sampling resolution.

   D. bandwidth.

Radio-Frequency Receivers

Radio-frequency (RF) *receivers* convert RF waves into the original message data sent by a distant wireless transmitter. In the simplistic sense, a receiver "undoes" everything that a transmitter does, in reverse sequence.

CHAPTER OBJECTIVES

In this chapter you will

- Compare direct-conversion and superheterodyne receivers.
- Learn the importance of sensitivity, selectivity, noise figure, and dynamic range in receiver design.
- Examine the individual stages of a single-conversion receiver.
- Discover how detectors work in various signal modes.
- Compare receivers for fast-scan, slow-scan, and digital television signals.
- Learn about receivers for specialized signal types.

Simple Designs

Simple circuits can work as RF receivers. Let's look at three examples of basic receiver schemes commonly employed in wireless electronics.

Diode-and-Amplifier Receiver

When we follow an RF diode with a sensitive class-A audio amplifier circuit, such as the one shown in Fig. 10-1, we get a receiver capable of picking up AM signals. The audio amplifier has enough power output to drive an earphone or headset.

The diode acts as a *detector*, also called a *demodulator*, to recover the modulating data from the RF signal. If we want the diode to perform this task properly, it must have low capacitance so that it functions as an RF rectifier but not as a capacitor. In this circuit, the polarity of the diode doesn't matter because the peak-to-peak received signal voltage is much less than the bias voltage on the JFET gate. We could reverse the diode's polarity and the circuit would still work.

In the circuit of Fig. 10-1, the variable capacitor and the inductor form a parallel-resonant circuit that we can use to frequency-adjust, or *tune*, the receiver to the desired signal. The position of the inductor tap matches

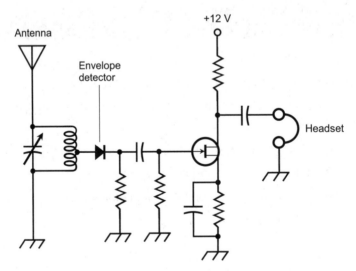

FIGURE 10-1 · A simple receiver for AM radio signals using a diode followed by an audio amplifier.

the antenna impedance to the input impedance of the amplifier. We must find the optimum position of the coil tap by experimentation; it depends on the choice of JFET, the value of the resistor between the diode and ground, and the type of antenna we connect to the system.

Direct-Conversion Receiver

A *direct-conversion receiver* combines incoming signals with the output of a variable-frequency *local oscillator* (LO). The received signal goes through at least one stage of weak-signal RF amplification and then into a *mixer* along with the output of the LO. The mixer also serves as the detector. Figure 10-2 is a simplified block diagram of a direct-conversion receiver.

For reception of on/off-keyed signals, we set the LO frequency a few hundred hertz above or below the signal frequency, so that the mixer produces an audio-frequency (AF) note at a *beat frequency* equal to the difference between the LO and signal frequencies. For reception of SSB signals, we must tune the LO precisely to the incoming signal's *suppressed-carrier frequency*.

TIP *A direct-conversion receiver doesn't always do a good job of differentiating between two signals whose frequencies lie close together because signals on either side of the LO frequency can interfere with one another. The audio amplifier can't distinguish between, say, an on/off-keyed Morse-code signal that's 750 Hz above the LO frequency and another similar signal that's 750 Hz below the LO frequency. Both signals produce AF output notes having the same pitch (750 Hz), so that they sound as if they're "right on top of each other."*

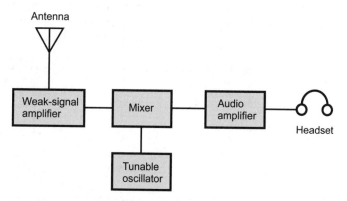

FIGURE 10-2 · Block diagram of a direct-conversion receiver.

Superheterodyne Receiver

A *superheterodyne receiver*, also called a *superhet*, uses one or more local oscillators and mixers to obtain an output signal that always stays at the same frequency. Therefore, all incoming signals can be processed using circuits optimized to work at a single frequency. A superhet can provide better gain, sensitivity, and selectivity than a direct-conversion receiver can.

In a typical superhet, the incoming signal from the antenna passes through a tunable, sensitive amplifier called the *front end*. The output of the front end mixes with the signal from a tunable LO. Either the sum or the difference signal, called the *first intermediate frequency* (or *first IF*), can be amplified and filtered to obtain high gain and excellent selectivity. If the first IF signal is detected, we have a *single-conversion superhet* because mixing takes place only once in the entire system (Fig. 10-3). Some superheterodyne receivers include a second mixer and second LO, converting the first IF signal to a lower-frequency *second IF* signal. We call this type of system a *double-conversion superhet*.

In any superheterodyne receiver, one or more sophisticated *IF bandpass filters* operate on a fixed frequency (the first IF, or the first and second IFs), resulting in excellent selectivity. Some bandpass filters allow for adjustable bandwidth. The gain and sensitivity of the superhet remain optimum as long as the IF amplifiers are all tuned to their proper frequencies. The process of tuning all

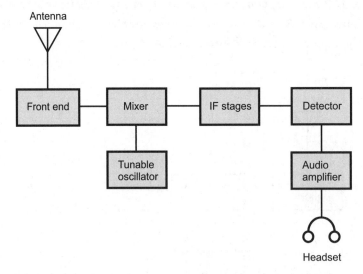

FIGURE 10-3 · Block diagram of a single-conversion superheterodyne receiver.

the IF stages to work together, and the maintenance of this ideal condition, is called *receiver alignment.*

TIP   *A superheterodyne receiver can intercept or generate "phantom" signals. We use the term* image *to describe a signal that originates outside the receiver at some frequency other than the one we want to hear, yet nevertheless passes through the mixer and IF amplifiers and shows up in the AF stages. If a harmonic of an LO output, or some spurious signal generated by an LO, shows up in a similar way, we call it a* birdie. *We can minimize problems with images and birdies by judiciously choosing the LO frequency or frequencies to minimize "conflicts."*

PROBLEM 10-1

Imagine that you have built the simple radio receiver diagrammed in Fig. 10-1 and are using it to listen to a local AM broadcast station by means of an outdoor wire antenna. What will happen if lightning strikes somewhere near your antenna, causing a transient "spike" that damages the diode so that it stops acting as a rectifier at RF?

SOLUTION

The diode will no longer separate the audio information from the incoming signal, but instead will appear as either a direct short circuit or as a tiny capacitor. As a result, no AF energy will appear in the headphones, and you won't hear anything. (If you were foolish enough to use the receiver during a lightning storm, you should be happy that the lightning strike damaged only the diode and didn't injure or kill you.) Signal diodes don't cost much money to replace, but you'll have to make sure that you get a new diode with the correct specifications—and never operate that receiver during a thunderstorm again.

The Modern Receiver

Wireless communications receivers operate over specific frequency ranges, known as *bands*, within the RF spectrum. The set of bands in which a particular receiver can operate depends on the application for which the receiver is designed and built.

Specifications

The *specifications* of a receiver indicate how well it can do what we want. The following are four major specifications that wireless-receiver engineers, technicians, and manufacturers often quote or discuss.

1. *Sensitivity:* This specification quantifies the number of microvolts (millionths of a volt, symbolized μV) that must exist at the antenna terminals to produce a certain *signal-to-noise ratio* (S/N) or *signal-plus-noise-to-noise ratio* (S+N/N) for the listener. Sensitivity depends to some extent on the gain of the front end, but the amount of noise the front-end transistor generates is more significant because subsequent stages amplify the front-end noise output as well as the signal output.

2. *Selectivity*: The width of a receiver passband (the range of signal frequencies allowed through the system at any given time) is established by a wideband *preselector* in the early RF amplification stages, and is narrowed down by filters in later amplifier stages. The preselector makes the receiver sensitive within a range of plus-or-minus (±) a few percent of the desired signal frequency. A *narrowband filter* at a later stage passes only the desired signal; signals in nearby channels are rejected.

3. *Noise figure*: This specification quantifies the amount of noise that a radio generates inside its own circuits (as opposed to noise coming from the outside environment). The less internal noise a receiver produces, in general, the lower is the noise figure, and the better is the S/N ratio. The noise figure attains increasing importance as the received frequency goes up, and becomes paramount at frequencies above about 30 MHz.

4. *Dynamic range*: This specification quantifies the extent to which a receiver can maintain a fairly constant output amplitude, and nevertheless, keep its rated sensitivity in the presence of signals ranging from very weak to extremely strong. In a receiver with poor dynamic range, weak signals might come in perfectly well if no strong signals exist in a nearby frequency, but the presence of a nearby strong signal will overwhelm the IF amplifier(s) and obliterate the weak signal.

Overview of the Single-Conversion System

Figure 10-3 illustrates the individual stages in a single-conversion superheterodyne radio receiver. Receiver designs can vary from this scheme, but Fig. 10-3 offers a good "generic" representation. The stages perform their respective tasks as follows.

- *Front end*: This stage comprises the first RF amplifier, and usually includes selective filters between the amplifier and the antenna. The dynamic range and sensitivity of a receiver are largely determined by the performance of the front end.

- *Mixer*: This circuit converts the variable signal frequency to a constant IF. The output frequency equals the sum or difference of the signal and LO frequencies.

- *IF stages*: Most of the amplification takes place in this part of the receiver. We can use narrowband filters to optimize the selectivity in these stages. Specialized filters and signal processors provide the desired bandwidth and frequency-response characteristics.

- *Detector*: This circuit extracts the data, also called *intelligence*, from the signal. The output of an ideal detector precisely duplicates the modulating signal from the original transmitting station.

- *Postdetector stages*: Following the detector, one or two stages of audio amplification boost the signal to a volume suitable for a speaker or headset. Alternatively, the signal can go to a printer, fax machine, video display, computer, or even a robot.

PROBLEM 10-2

Imagine a superheterodyne receiver built according to the design shown in Fig. 10-3. What will happen if the local (tunable) oscillator frequency becomes unstable?

SOLUTION

Assuming that the intended received signal has a constant frequency, any change in the LO frequency will produce a similar change in the sum or difference frequency at the output of the mixer. As a result, the signal frequency will fluctuate in the IF stages, making effective reception difficult or impossible.

Predetector Stages

In any RF receiver, the stages preceding the detector must provide reasonable gain while producing as little internal noise as possible. They should also offer some selectivity, so that signals at frequencies far removed from the desired one don't "leak through" and produce troublesome images in circuits farther on down the line.

Preamplifier

If we want to receive especially weak signals, we can place a self-contained RF *preamplifier* between a receiver and its antenna. Figure 10-4 shows a tunable RF preamplifier using an N-channel JFET. An *LC* circuit at the antenna circuit provides some selectivity. This amplifier can generate 5 to 10 dB of useful gain, depending on the frequency and the choice of FET. A receiving preamplifier must exhibit excellent linearity for the signal waveform, mandating the use of the class-A mode. Nonlinearity results in *intermodulation distortion* (IMD) that increases the risk of images, and can also reduce the dynamic range.

Front End

At low and medium frequencies, we often observe substantial atmospheric noise caused by lightning and other environmental disturbances. Above 30 MHz, atmospheric noise diminishes; then the main factor that limits receiver sensitivity is internal noise (electrical noise generated within the circuits themselves). For this reason, front-end design becomes increasingly critical as the frequency rises, especially above 30 MHz.

The front end, like a preamplifier, must be as linear as possible. The greater the degree of nonlinearity, the greater the risk of IMD and all of its attendant problems. The front end should offer the greatest possible dynamic range.

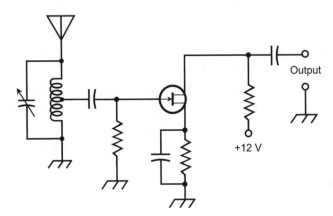

FIGURE 10-4 · A tunable, stand-alone RF preamplifier for use with a radio receiver.

Otherwise a strong local signal can cause the receiver to *desensitize* (lose sensitivity) over much or all of its operating frequency range.

Preselector

The *preselector* provides a bandpass response, or degree of selectivity, that improves the S/N ratio and reduces the likelihood of receiver desensitization by a strong signal far removed from the operating frequency. In a superheterodyne receiver, the preselector helps the system reject image signals. We can tune a preselector by means of *tracking* with the receiver's main tuning control so that the preselector remains correctly adjusted as we move from one frequency to another.

TIP   *Some older receivers, particularly "shortwave radios" dating from the 1950s or earlier, incorporate preselectors that the operator can adjust independently of the main tuning control. The operator must fine-tune, or "tweak," the preselector after making a major change to the receiver's main tuning control.*

PROBLEM 10-3

Suppose that a receiver has a preselector that tracks along with the main tuning control. What will happen if such a receiver gets out of alignment so that the passband of the preselector doesn't always coincide with the frequency of the desired incoming signal?

SOLUTION

When *misalignment* of this sort happens, the overall gain of the receiver decreases at some frequencies. If the misalignment gets severe, the sensitivity and dynamic range also suffer.

Mixers

A mixer requires a nonlinear, waveform-distorting component or circuit that makes it easy for the signals to beat against each other and produce energy at the sum and difference frequencies. The new signals are known as *mixing products*. When we use nonamplifying elements, such as diodes, to mix signals, we have a *passive mixer*. The term "passive" comes from the fact that the circuit doesn't need an external source of power. As such, it provides no signal gain (and in fact introduces some loss). An *active mixer* employs one or more transistors or integrated circuits to obtain signal gain. The amplifying device is usually biased in class-AB$_1$, class-AB$_2$, or class-B.

TIP *When an engineer designs a single-conversion receiver, she can tune the mixer output to the optimum response at either the sum frequency or the difference frequency, as desired.*

IF Stages

A high IF (several megahertz) works better than a low IF (less than 1 MHz) when we want to attenuate unwanted image signals, a process called *image rejection*. However, a low IF offers better performance when we want to obtain a high degree of selectivity (a narrow response bandwidth). A double-conversion receiver has a comparatively high first IF and a low second IF to offer us the best features of both technologies.

Two or more IF amplifiers can be *cascaded*, or connected one after another, to obtain high gain and high selectivity at each IF. We can use transformer coupling between individual amplifiers to facilitate *LC* tuning of each stage. These amplifiers come after the mixer stage and before the detector stage. Double-conversion receivers have two sets, called *chains*, of IF amplifiers. The *first IF chain* follows the first mixer and precedes the second mixer. The *second IF chain* follows the second mixer and precedes the detector.

We can quantify the selectivity of an IF chain by comparing the bandwidths for two power-attenuation values of an incoming signal, usually 3 dB down and 30 dB down with respect to the gain obtained when the signal appears at the center of the passband. The ratio of the 30-dB selectivity to the 3-dB selectivity is called the *shape factor*. In some receivers, engineers specify voltage-attenuation values instead of power-attenuation values, citing the ratio of the 60-dB selectivity to the 6-dB selectivity.

Still Struggling

As the shape factor decreases, the IF gain goes down with increasing rapidity, as a signal moves from within the passband to outside it. A small shape factor, such as 2:1 or less, results in excellent rejection of unwanted signals, while allowing all of the desired signals to pass through with little or no attenuation.

Detectors

Engineers define *detection* as the recovery of intelligence, such as audio, images, or printed data, from a received RF signal. In some text, the term *demodulation* is used instead.

Detection of AM

We can extract the modulating waveform from an AM signal by rectifying that signal. Figure 10-5AB (upper diagram) is a simplified graphical diagram of this process. The rapid pulsations occur at the carrier frequency. The slower fluctuation (dashed line) comes out as a duplication of the modulating-intelligence waveform. The carrier pulsations are smoothed out by passing the output through a capacitor large enough to hold the charge for one carrier current cycle, but not so large that it "levels out" the cycles of the modulating intelligence. We call this scheme *envelope detection*.

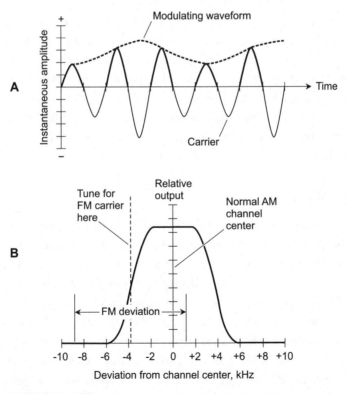

FIGURE 10-5AB · At A, envelope detection of AM, shown in the time domain. At B, slope detection of FM, shown in the frequency domain.

Still Struggling

Here's an alternative description of what happens in AM demodulation. Any AM signal contains a carrier (an RF sine wave) and sidebands (signals at frequencies slightly above and below that of the carrier). The sidebands arise in the transmitted signal when the carrier mixes with voice, music, or image data signals. An AM envelope detector separates the sidebands from the carrier, which contains energy but no information. A single shunt capacitor immediately following the detector can filter out the residual, now-useless carrier from the demodulated signal. The resulting output replicates the original transmitted data waveform.

Detection of On/Off Keying

When we want to demodulate an on/off keyed signal, we can inject a local, unmodulated RF carrier into the receiver a few hundred hertz above or below the frequency of the incoming signal. The injected signal comes from a tunable *beat-frequency oscillator* (BFO). The BFO signal and the desired signal mix to produce an audio note at the difference frequency. The operator can fine-tune the BFO to a frequency that results in a comfortable listening pitch, usually 400 to 1000 Hz above or below the main signal frequency. We call this process *heterodyne detection*.

Detection of FSK

We can demodulate frequency-shift keyed (FSK) signals using the same heterodyne-detection technique that we use for on/off-keyed signals. The received-signal carrier beats against the BFO signal in the mixer, producing an audio note that alternates between two different pitches. We set the BFO frequency a few hundred hertz above or below both the mark and space FSK-signal frequencies. The *frequency offset*, or difference between the BFO and FSK-signal frequencies, determines the audio output frequencies. We must set the BFO at exactly the correct frequency, so the audio output tones occur at the correct pitches. Otherwise the output device, such as a personal computer or teleprinter, won't properly sense the detector output.

Detection of FM

Frequency-modulated (FM) signals can be detected in various ways. As we might expect, the most effective methods require the most complex and expensive circuits. Four common systems for detection of FM exist, as follows.

1. *Slope detection:* An AM receiver can demodulate FM signals if we set the receiver frequency near, but not on, the FM unmodulated-carrier frequency. An AM receiver has a narrowband filter with a passband of a few kilohertz. This feature provides a selectivity curve, such as the one shown in Fig. 10-5AB (lower diagram). If the FM unmodulated-carrier frequency lies in the middle of the filter-response *skirt* (sloping part of the curve), frequency modulation makes the signal move in and out of the passband, causing the instantaneous receiver output volume to vary.

2. *Phase-locked loop* (PLL): If we inject an FM signal into a PLL, the loop produces a fluctuating error voltage that duplicates the modulating signal waveform. A *limiter* can be placed ahead of the PLL so that the signal passes through the limiter before it gets to the PLL. The limiter keeps the receiver from responding to variations in the signal amplitude. Weak signals "break up" rather than fade in an FM receiver that employs limiting. A marginal FM signal sounds like the signal from a cell phone at the fringe of the coverage zone.

3. *Discriminator:* This type of FM detector produces an output voltage that depends on the instantaneous signal frequency. When the signal is at the center of the passband, the output voltage equals zero. If the instantaneous signal frequency falls below center, the output voltage becomes positive. If the frequency rises above center, the output becomes negative. A discriminator responds to amplitude variations as well as to frequency variations, but we can overcome this potential problem by placing a limiter stage ahead of the discriminator circuit.

4. *Ratio detector:* This type of FM detector comprises a discriminator with a built-in limiter. The original design was developed by RCA (Radio Corporation of America). Figure 10-5C shows a simple ratio detector circuit. The "balance" potentiometer should be set by a lab technician for the best received signal quality under controlled conditions. The excellent limiting characteristics of this circuit enhance the immunity of the FM receiver to external noise sources, such as thunderstorms, internal combustion engines, and certain home appliances.

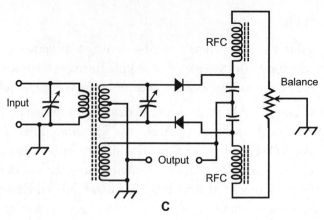

FIGURE 10-5C · At C, a ratio detector circuit for demodulating FM signals.

Detection of SSB

A single-sideband (SSB) signal constitutes an AM signal with the carrier and one of the sidebands removed. We can combine an incoming SSB signal with the signal from an unmodulated LO in a nonlinear circuit, reproducing the original modulating data from the source transmitter. We call this process *product detection*. It takes place at a single frequency, rather than at a variable frequency as in direct-conversion reception.

Figure 10-5DE shows passive and active product detectors, respectively. The circuit at D uses diodes and offers no amplification. The circuit at E employs a transistor to provide some signal gain. The essential characteristic of either circuit is nonlinearity in the semiconductor device, encouraging the generation of sum-frequency and difference-frequency signals. In this respect, product detectors act as specialized mixers.

PROBLEM 10-4

Suppose that the transistor in the circuit of Fig. 10-5E suddenly stops amplifying, but nevertheless allows both the signal input and the LO input to pass through, and still mixes those signals to produce product detection. What will happen at the output?

SOLUTION

The output signal will contain the same data as it did before the amplification failure, but at a much weaker amplitude.

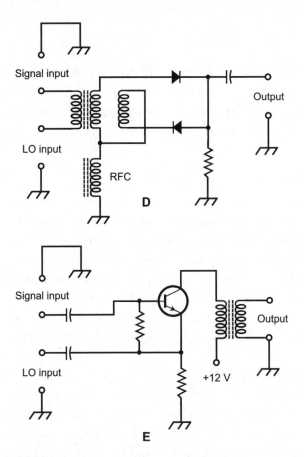

FIGURE 10-5DE · At D, a product detector using diodes. At E, a product detector using an NPN bipolar transistor biased as a class-B amplifier.

PROBLEM 10-5

Imagine that, in the circuit of Fig. 10-5E, we want to use a PNP bipolar transistor instead of the NPN device. Suppose that the characteristic curves of the PNP device are the same as those for the NPN device. In what way must we alter the circuit to make it work with the PNP transistor?

SOLUTION

We must reverse the polarity of the power-supply voltage at the terminal marked +12V. To reflect this change in Fig. 10-5E, we need do nothing more than relabel the power-supply terminal so that it specifies −12V (or something close to it).

Audio Stages

In a radio receiver, enhanced selectivity can be obtained by tailoring the AF response in stages following the detector. The audio stages can also provide extra amplification. If we want to use a loudspeaker, the audio stages must provide enough output power to drive it (at least 1 W).

Audio Filtering

A voice signal requires a band of frequencies ranging from about 300 to 3000 Hz in order to come out understandable to a listener. An *audio bandpass filter* with a passband of 300 to 3000 Hz (which represents a bandwidth of 2700 Hz) can improve the quality of reception with some voice receivers. An ideal voice audio filter has little or no attenuation within the passband range and high attenuation outside the range. The skirts (dropoffs in the selectivity curve) are steep. If we graph the attenuation of such a filter versus the frequency, we get a so-called *rectangular response*, characterized by a flat "top" and nearly vertical "sides."

An on/off-keyed signal requires only a small amount of bandwidth for optimum intelligibility. Audio filters can narrow the response bandwidth to as little as 100 Hz. Passbands narrower than 100 Hz produce *ringing*, degrading the quality of reception. The center frequency of the audio filter should be set somewhere between about 400 and 1000 Hz, which represents a comfortable listening pitch for most people. Most *radiotelegraphy* (Morse-code) audio filters, popular among amateur radio operators, have adjustable center frequencies.

An *audio notch filter* is a band-rejection filter with a sharp, narrow response. An interfering carrier, called a *heterodyne*, produces an audio note of constant frequency in the receiver output. A notch filter can get rid of the annoying heterodyne note without adversely affecting the rest of the passband. A good audio notch filter can be fine-tuned over the entire AF range, roughly 20 Hz to 20 kHz. Audio notch filters for communications use have narrower ranges, such as 300 to 3000 Hz.

TIP *Some audio notch filters tune themselves automatically. When a heterodyne appears and remains for a few hundredths of a second, a built-in* **microcontroller** *("computer on a chip") causes the notch to center itself on the heterodyne frequency. You can listen to the heterodyne "die off" shortly after it appears.*

Squelching

A *squelch* silences a receiver when no signal comes in, allowing clear reception of signals whenever they appear. Most FM communications receivers use squelching systems. The squelch is normally *closed*, silencing the receiver when no signal exists. A signal *opens* the squelch when its amplitude exceeds the *squelch threshold*, which the operator can adjust at will. When the squelch is open, all incoming signals (as well as any accompanying noise or interference) within the passband can reach the receiver output.

In some systems, the squelch will not open unless the signal has certain characteristics. The most common method of so-called *selective squelching* uses AF tone generators. This precaution can prevent unauthorized transmitters from actuating receivers. The squelch opens only for signals accompanied by a *subaudible note* (below 300 Hz) having the correct frequency, or by a rapid sequence of brief audible notes having the correct frequencies in a specific, predetermined order at the beginning of every incoming transmission.

PROBLEM 10-6

Suppose that a receiver's audio output circuits can pass a band of frequencies much greater than necessary to reproduce a voice. For example, suppose the circuits can pass frequencies from 20 Hz to 20 kHz. If we listen to a voice communications signal with such a receiver, what should we expect?

SOLUTION

The receiver will work perfectly well if minimal noise exists, and as long as no signals occupy nearby frequencies. But we'll experience severe interference if signals appear within 10 or 20 kHz of our frequency. In addition, if much external noise such as thunderstorm "static" exists, more of this noise will get through along with the essential voice information, compared with a circuit that passes energy only in the standard voice communications range of 300 to 3000 Hz.

Television Reception

A television (TV) receiver has a tunable front end, an oscillator and mixer, a set of IF amplifiers, separate detectors for the video and audio portions of the signal, a display with associated peripheral circuitry, and a loudspeaker.

Fast-Scan TV

Figure 10-6 is a simplified block diagram of a receiver for the now-outdated analog *fast-scan television* (FSTV) mode, which some "ham" radio operators still experiment with. The signal has 525 lines per frame and conveys 30 complete frames per second. As of this writing, the analog FSTV mode, also called NTSC (for National Television Systems Committee) TV, has been replaced by digital TV for commercial broadcasting. Television broadcasts are made on various channels.

Digital Television

Digital television constitutes the transmission and reception of moving video images in digitized form. Communications satellites have proven popular for this purpose.

Until the early 1990s, a satellite television installation required a dish antenna 6 to 10 ft (roughly 2 to 3 m) in diameter. Digitization has changed this situation. In any communications system, digitization allows the use of smaller receiving antennas, smaller transmitting antennas, and/or lower transmitter power levels. Engineers have managed to get the diameter of the receiving dish down to about 2 ft (60 cm or so).

The *Radio Corporation of America* (RCA) pioneered digital satellite TV with its so-called *Digital Satellite System* (DSS). The analog signal was changed into digital pulses at the transmitting station using an *analog-to-digital converter* (A/D

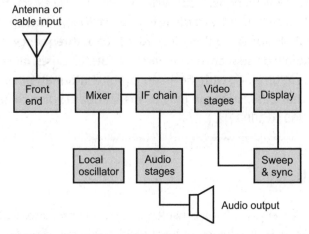

FIGURE 10-6 · Block diagram of a fast-scan television (FSTV) receiver.

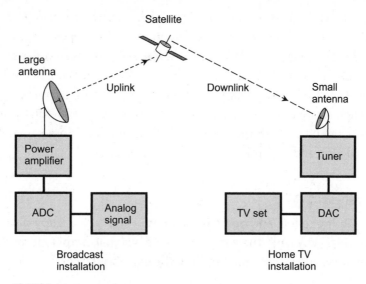

FIGURE 10-7 · A complete digital TV link.

converter or ADC). The digital signal was amplified and sent up to a satellite. The satellite had a *transponder* that received the signal, converted it to a different frequency, and retransmitted it back to the earth. A portable dish picked up the downcoming (or *downlink*) signal.

Although digital satellite TV technology has evolved since the initial days of the RCA DSS, today's systems work in essentially the same way as the original one did. A *tuner* circuit selects the channel that the subscriber wants to view. The digital signal is amplified and processed. It can then be viewed on a digital TV receiver, or changed into analog form suitable for viewing on an analog TV set by means of a *digital-to-analog converter* (D/A converter or DAC), as shown in Figure 10-7. Alternatively, a special high-definition receiving set can be used.

TIP *You can obtain computer software that allows digital TV signals to be viewed on a personal computer. The image detail, or **resolution**, is superior to that of NTSC TV. You can also watch and hear some digital TV broadcasts over the Internet if you have an exceptional connection, a good video display, and good speakers.*

Slow-Scan TV

A *slow-scan television* (SSTV) communications station needs a *transceiver* (transmitter/receiver combination) with SSB capability, a standard NTSC TV set or a computer with the appropriate software, a video camera, and a *scan converter*. The standard transmission rate is 8 frames per second. This reduces the required bandwidth to the essential audio voice frequency range of 300 Hz to 3000 Hz. That way, the SSTV signal can be sent directly over conventional SSB equipment without modification. The scan converter consists of two data converters (one for receiving and the other for transmitting), a memory circuit, a tone generator, and a detector. Scan converters are commercially available. Computers can be programmed to perform this function. Some amateur radio operators design and build their own scan converters.

PROBLEM 10-7

Why do the uplink and downlink signals in Fig. 10-7 occupy different frequencies? Why can't the satellite receive a signal, amplify it on the same frequency, and retransmit it back to the earth's surface?

SOLUTION

The downlink signal, as it emerges from the satellite's transmitting antenna, must not interfere with the received uplink signal arriving from the earth and reaching the receiving antenna. If the two signals were on the same frequency, this interference would be practically impossible to prevent because of the close proximity of the two antennas. Even when the downlink frequency differs significantly from the uplink frequency, we need *isolation circuits* to prevent the transmitted signal from desensitizing the receiver. In addition, the satellite receiver's frequency must never lie on, or near, any harmonic of the satellite transmitter's frequency.

Specialized Modes

Some exotic wireless communications techniques can prove effective under difficult circumstances. Let's look at a few examples.

Dual-Diversity Reception

Dual-diversity reception, also known simply as *diversity reception*, reduces the adverse effects of *fading* (fluctuations in received signal strength) in radio reception that occurs when signals from distant transmitters return to the earth from the ionosphere. This phenomenon often takes place at frequencies from 3 to 30 MHz, a range sometimes called the *shortwave band*. Two receivers are used, both tuned to the same signal. They employ separate antennas, spaced several wavelengths apart. We combine the outputs of the receivers, in phase coincidence, at the input of a single audio amplifier, as shown in Fig. 10-8. We must set both receivers to precisely the same frequency, so that they both respond to the same incoming signal.

TIP *Some diversity-reception installations employ three or more antennas and receivers. This technique provides even better immunity to fading than dual-diversity reception does, but it compounds the tuning difficulty and increases the expense.*

Synchronized Communications

Digital signals require less bandwidth than analog signals to convey a given amount of information per unit of time. In *synchronized communications*, also known as *coherent communications*, the transmitter and receiver operate from a common time standard to optimize the amount of data that we can send within a particular frequency channel or band. The receiver and transmitter operate in "lock-step." The receiver evaluates each transmitted bit as a unit, for a block of time lasting from the bit's exact start to its exact finish. This process allows the use of a receiving filter having extremely narrow bandwidth.

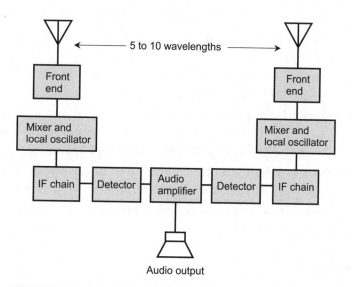

FIGURE 10-8 · Dual-diversity reception uses two antennas and two identical radio receivers tuned to the same frequency.

Rigorous transmitter/receiver synchronization requires an external *time standard*, such as the broadcasts of the shortwave radio station WWV. *Frequency dividers* derive the necessary synchronizing frequencies from the standard. A tone or pulse is generated in the receiver output for a particular bit if, but only if, the average signal voltage exceeds a certain value over the duration of that bit. False signals, such as might result from filter ringing, *sferics* (atmospheric "static" from thunderstorms), or other noise, are ignored because they do not produce sufficient average bit voltage.

TIP   *Experiments with synchronized communications have shown that the improvement in S/N ratio, compared with nonsynchronized systems, can range up to about 10 dB at low to moderate data speeds.*

Digital Signal Processing (DSP)

Digital signal processing (DSP) can extend the workable range of a communications circuit because it allows reception under worse conditions than we could ever hope for without it. Digital signal processing also improves the quality of fair signals, so that the receiving equipment or operator makes fewer errors. In circuits that use only digital modes, DSP can "clean up" the signal, making it possible to replicate digital data over and over again (that is, to produce *multigeneration copies*).

When we use DSP with analog modes, such as SSB or SSTV, the circuits change the signals from analog form into digital form using A/D conversion. Then the digital data is "cleaned up" so the pulse timing and amplitude adhere strictly to the set of technical standards (known as the *protocol*) for the type of digital data involved. Finally, the digital signal is changed back to the original voice or video using D/A conversion.

A system for DSP minimizes noise and interference in a received digital signal, as shown in Fig. 10-9. At A, we see a hypothetical signal before DSP; at B, we see the same signal after DSP. If the incoming signal amplitude remains or averages above a certain level for an interval of time, then the DSP system generates a *high* (logic 1) output signal. If the incoming signal amplitude remains or averages below the critical threshold for a time interval, then the output is *low* (logic 0).

Multiplexing

Signals in a communications channel or band can be intertwined in various ways. The most common methods are *frequency-division multiplexing* (FDM)

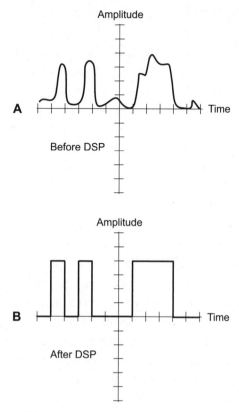

FIGURE 10-9 · Digital signal processing (DSP) can "clean up" a signal.

and *time-division multiplexing* (TDM). Multiplexing requires an *encoder* at the transmitter and a *decoder* at the receiver. In FDM, the channel is broken down into subchannels. The carrier frequencies of the signals are spaced so they do not overlap. In TDM, signals are broken into segments by time, and then the segments are transferred in a rotating sequence. The receiver must be synchronized with the transmitter. They can both be clocked from a time standard such as WWV. In both the FDM mode and the TDM mode, each signal remains entirely independent of all the others.

Spread-Spectrum Communications

In *spread-spectrum communications*, the main carrier frequency is rapidly varied independently of signal modulation, and the receiver is programmed to follow. As a result, the probability of *catastrophic interference*, in which a single strong interfering signal can obliterate the desired signal, vanishes to near zero. In addition, unauthorized people will find it practically impossible to eavesdrop on a communication in progress.

Frequency-spreading functions can be complex indeed. If the operators of the transmitting station and all receiving stations do not divulge the function to anyone else, and if they do not tell anyone else about the existence of their communication, then no one else on the band will know that the contact is taking place. Any number of receivers can intercept a spread-spectrum signal as long as the receiving operators know the frequency-spreading function. The use of such a function constitutes a form of *data encryption*.

A common method of generating spread-spectrum signals goes by the descriptive name *frequency hopping*. The transmitter has a list of channels that it follows in a certain order. The receiver is programmed with this same list, in the same order, so that its frequency follows along exactly with that of the transmitter. The *dwell time* equals the length of time that the transmitter and receiver remain on any given frequency between hops. Numerous *dwell frequencies* exist to "dilute" the signal energy so that, if someone tunes to any particular fixed frequency in the sequence, the signal will pass unnoticed.

Another method of obtaining spread-spectrum communications, called *frequency sweeping*, involves frequency-modulating the main transmitted carrier with a waveform that guides it up and down over the assigned band. This FM function operates independently from the signal intelligence waveform. A receiver can intercept a frequency-swept signal if, but only if, its tuning varies according to the same function, over the same band, at the same frequency, and in the same phase as the transmitter tuning varies.

Still Struggling

You might wonder whether or not spread-spectrum technology allows for infinitely many signals to "fit into" a band of finite frequency width. The answer is no, we can't do that. As a band becomes occupied with an increasing number of spread-spectrum signals, the overall noise level in the band appears to increase, as "heard" by any receiver in the band. A practical limit, therefore, exists as to the number of spread-spectrum contacts that a band can handle. This limit, in terms of the number of signals, is about the same as it would be if all the signals constituted conventional, constant-frequency emissions.

PROBLEM 10-8

What will happen if the timing between the transmitter and receiver gets disrupted in a synchronized communications system?

SOLUTION

The enhanced S/N ratio will be lost. In some systems, if synchronization fails, the system will revert to a standard, nonsynchronized mode (called *asynchronous communications*), so reception will still take place, although at a reduced level of performance. In less sophisticated systems, complete failure occurs because the receiver "gets lost" and can no longer find the transmitted signal.

QUIZ

Refer to the text in this chapter if necessary. A good score is eight correct. The answers are listed in the back of the book.

1. Figure 10-10 is a schematic diagram of a specialized circuit with one serious error. Which of the following single actions can we take to correct that error?
   A. Remove the resistor between the JFET source and ground.
   B. Replace the gate resistor with a capacitor.
   C. Connect the non-JFET end of the drain resistor to −12 V DC instead of ground.
   D. Replace the P-channel JFET with an NPN bipolar transistor.

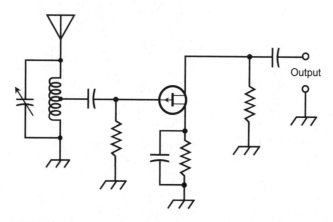

FIGURE 10-10 · Illustration for Quiz Questions 1 and 2.

2. Once we have made the necessary correction to the circuit shown in Fig. 10-10, what useful role can it play in a communications receiving system?

   A. AF amplifier
   B. RF preamplifier
   C. RF envelope detector
   D. RF mixer

3. In a voice communications system, we can keep a receiver from "hearing" unauthorized transmissions by means of

   A. a phase-locked loop.
   B. selective squelching.
   C. envelope detection.
   D. product detection.

4. If we want to receive an SSB signal using a direct-conversion receiver, we should set the LO frequency

   A. a few hundred hertz away from the signal's suppressed-carrier frequency.
   B. precisely at the signal's suppressed-carrier frequency.
   C. at a harmonic of the suppressed-carrier frequency.
   D. in the middle of the sideband containing the voice signal energy.

5. The circuit shown in Fig. 10-11 can serve as a detector for

   A. SSB signals.
   B. on/off-keyed signals.
   C. FSK signals.
   D. All of the above

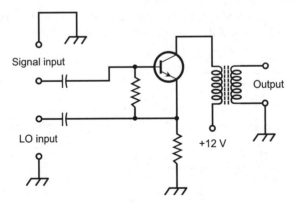

FIGURE 10-11 • Illustration for Quiz Questions 5 and 6.

6. **If we want to get the best possible results from the circuit of Fig. 10-11, the amplifying device (in this case an NPN transistor) should exhibit**
    A. high power output.
    B. a small shape factor.
    C. broad bandwidth.
    D. nonlinearity.

7. **We can improve digital-signal reception under marginal conditions (such as a weak signal or a high noise level) by taking advantage of**
    A. digital signal processing.
    B. spread-spectrum technology.
    C. selective squelching.
    D. a passive mixer.

8. **In a well-designed superheterodyne communications receiver, the signal from the antenna goes directly to the**
    A. detector.
    B. first IF.
    C. first mixer.
    D. front end.

9. **We can break up several data signals into segments by time, and send all of the signals on a single carrier in a rotating sequence, to accomplish**
    A. intermodulation.
    B. mixing.
    C. multiplexing.
    D. product detection.

10. **A preselector in the RF stages of a receiver**
    A. attenuates signals at frequencies far removed from the desired frequency.
    B. prevents birdies from reaching the front end.
    C. optimizes the shape factor in the IF stages.
    D. prevents the detector from breaking into oscillation.

chapter **11**

Telecommunications

The term *telecommunications* refers to the electronic transfer of information between individuals, businesses, and/or governments. This activity can take place over *landline* (hard-wired systems), by *wireless* (EM waves), or by a combination of landline and wireless modes.

CHAPTER OBJECTIVES

In this chapter you will

- Learn how machines "count" using various numeration schemes.
- Compare units of data storage and transfer speed.
- Outline the fundamental elements of a communications network.
- See how satellite links work, and compare different satellite types.
- Learn the basics of personal wireless networking.
- Discover how engineers maintain and enhance network security.

Numeration Systems

In everyday life, most of us deal with the *decimal number system*, which makes use of digits from the set

$$D = \{0, 1, 2, 3, 4, 5, 6, 7, 8, 9\}$$

Machines such as computers and communications devices can, and generally do, take advantage of other numeration systems.

Decimal

The familiar *decimal number system* is also called *base 10* or *radix 10*. When we express nonnegative integers (whole numbers) in this system, we multiply the right-most digit by 10^0, or 1. We multiply the next digit to the left by 10^1, or 10. The power of 10 increases as we move to the left. Once we've done the digit multiplication, we add up the values. For example, the decimal quantity 862,058 breaks down as follows:

$$8 \times 10^0 + 5 \times 10^1 + 0 \times 10^2$$
$$+ 2 \times 10^3 + 6 \times 10^4 + 8 \times 10^5$$
$$= 862,058$$

Binary

The *binary number system* provides a means of expressing numbers using only the digits 0 and 1. Some mathematicians call this system *base 2* or *radix 2*. When we express nonnegative integers in binary notation, we multiply the rightmost digit by 2^0, or 1. The next digit to the left is multiplied by 2^1, or 2. The power of 2 increases as we continue to the left, so we get a "fours" digit, then an "eights" digit, then a "16s" digit, and so on. For example, consider the decimal number 94. In the binary system, we would write this quantity as 1011110. It breaks down as follows:

$$0 \times 2^0 + 1 \times 2^1 + 1 \times 2^2 + 1 \times 2^3$$
$$+ 1 \times 2^4 + 0 \times 2^5 + 1 \times 2^6$$
$$= 94$$

Octal

Another scheme, sometimes used in computer programming, goes by the name *octal number system* because it has eight symbols (according to our way of thinking), or 2^3. Every digit constitutes an element of the set

$$O = \{0, 1, 2, 3, 4, 5, 6, 7\}$$

Some people call this system *base 8* or *radix 8*. When we express nonnegative integers in octal notation, we multiply the rightmost digit by 8^0, or 1. The next digit to the left is multiplied by 8^1, or 8. The power of 8 increases as we continue to the left, so we get a "64s" digit, then a "512s" digit, and so on. For example, we render the decimal quantity 3085 in octal form as 6015. We can break it down into the following sum:

$$5 \times 8^0 + 1 \times 8^1 + 0 \times 8^2 + 6 \times 8^3$$

$$= 3085$$

Hexadecimal

Computer programmers often work in the *hexadecimal number system*. It has 16 (2^4) symbols: the usual 0 through 9 plus six more, represented by the uppercase English letters A through F, yielding the digit set

$$H = \{0, 1, 2, 3, 4, 5, 6, 7, 8, 9, A, B, C, D, E, F\}$$

This system is sometimes called *base 16* or *radix 16*. All of the hexadecimal digits 0 through 9 represent the same values as their decimal counterparts. However, we have the following additional digits:

- Hexadecimal A equals decimal 10
- Hexadecimal B equals decimal 11
- Hexadecimal C equals decimal 12
- Hexadecimal D equals decimal 13
- Hexadecimal E equals decimal 14
- Hexadecimal F equals decimal 15

When we express nonnegative integers in hexadecimal notation, we multiply the rightmost digit by 16^0, or 1. We multiply the next digit to the left by 16^1, or 16. The power of 16 increases as we continue to the left, so we get a "256s" digit, then a "4096s" digit, and so on. For example, we would write the decimal quantity 35,898 in hexadecimal form as 8C3A. Remembering that $C = 12$ and $A = 10$, we can break the hexadecimal number down as follows:

$$A \times 16^0 + 3 \times 16^1 + C \times 16^2 + 8 \times 16^3$$

$$= 35,898$$

PROBLEM 11-1

Express the binary number 10011011 in decimal form.

SOLUTION

We can add the digits up by going from left to right, or from right to left. It doesn't matter which way we go, but most people find it easier to keep track of the progression by starting at the "ones" place and working toward the left from there. The digits add up as follows:

$$1 \times 2^0$$
$$1 \times 2^1$$
$$0 \times 2^2$$
$$1 \times 2^3$$
$$1 \times 2^4$$
$$0 \times 2^5$$
$$0 \times 2^6$$
$$1 \times 2^7$$
$$\overline{155}$$

PROBLEM 11-2

Express the decimal number 1,000,000 in hexadecimal form. Warning: This problem involves some tedious arithmetic!

SOLUTION

The values of the digits in a whole (nonfractional) hexadecimal number, proceeding from right to left, constitute ascending nonnegative integer powers of 16. Therefore, a whole hexadecimal number n_{16} has the form

$$n_{16} = \ldots + (f \times 16^5) + (e \times 16^4) + (d \times 16^3)$$
$$+ (c \times 16^2) + (b \times 16^1) + (a \times 16^0)$$

where $a, b, c, d, e, f, \ldots$ represent single-digit hexadecimal numbers from the set

$$H = \{0, 1, 2, 3, 4, 5, 6, 7, 8, 9, A, B, C, D, E, F\}$$

Let's begin by finding the largest power of 16 that's less than or equal to 1,000,000. This value is $16^4 = 65,536$. We divide 1,000,000 by 65,536, obtaining 15 plus a remainder. We represent the decimal 15 as the hexadecimal F.

Now we know that the hexadecimal expression of the decimal number 1,000,000 has the form

$$(F \times 16^4) + (d \times 16^3) + (c \times 16^2) + (b \times 16^1) + a$$
$$= Fdcba$$

To find the value of d, we note that

$$15 \times 16^4 = 983,040$$

This quantity is 16,960 smaller than 1,000,000, so we must find the hexadecimal equivalent of decimal 16,960 and add it to hexadecimal F0000. The largest power of 16 that's less than or equal to 16,960 is 16^3, or 4096. We divide 16,960 by 4,096 to obtain 4 plus a remainder. Now we know that $d = 4$ in the above expression, so the decimal 1,000,000 has the hexadecimal form

$$(F \times 16^4) + (4 \times 16^3) + (c \times 16^2) + (b \times 16^1) + a$$
$$= F4cba$$

To find the value of c, we note that

$$(F \times 16^4) + (4 \times 16^3) = 983,040 + 16,384$$
$$= 999,424$$

This quantity is 576 smaller than 1,000,000, so we must find the hexadecimal equivalent of decimal 576 and add it to hexadecimal F4000. The largest power of 16 that's less than or equal to 576 is 16^2, or 256. We divide 576 by 256 to get 2 plus a remainder. Now we know that $c = 2$ in the above expression, so the decimal 1,000,000 is equivalent to the hexadecimal

$$(F \times 16^4) + (4 \times 16^3) + (2 \times 16^2) + (b \times 16^1) + a$$
$$= F42ba$$

To find the value of b, we note that

$$(F \times 16^4) + (4 \times 16^3) + (2 \times 16^2) = 983,040 + 16,384 + 512$$
$$= 999,936$$

This quantity is 64 smaller than 1,000,000, so we must find the hexadecimal equivalent of decimal 64 and add it to hexadecimal F4200. The largest power of 16 that's less than or equal to 64 is 16^1, or 16. We divide 64 by 16, obtaining 4 without any remainder. Now we know that $b = 4$ in the above expression, so the decimal 1,000,000 is equivalent to the hexadecimal

$$(F \times 16^4) + (4 \times 16^3) + (2 \times 16^2) + (4 \times 16^1) + a$$
$$= F424a$$

When we found b, we were left with no remainder. Therefore, all the digits to the right of b (in this case, that means only the digit a) must equal 0. We've gone through a tedious process, but we've determined that the decimal number 1,000,000 equals the hexadecimal number F4240. To check our work, we can do the arithmetic the other way, converting the hexadecimal number to decimal form and multiplying out the digits from right to left, as follows:

$$0 \times 16^0$$
$$4 \times 16^1$$
$$2 \times 16^2$$
$$4 \times 16^3$$
$$\underline{15 \times 16^4}$$
$$1,000,000$$

Binary Communications

The use of *binary data* (digital information transmitted using only two different states) provides excellent communications accuracy and efficiency. If we want to attain *multilevel signaling* (digital transmission with more than two states), then we can represent each and every signal level with a specific group of binary digits, representing a unique binary number. A group of three binary digits, for example, can represent up to 2^3 (that is, eight) levels. A group of four binary digits can represent up to 2^4 (that is, 16) levels. Engineers and technicians combine and abbreviate the words *binary digit* to obtain the single word *bit*.

Forms of Binary Signaling

In communications systems, binary data has less sensitivity to noise and interference than analog data does. For this reason, binary systems usually work better than analog systems, producing fewer errors and allowing communication under more adverse circumstances. Since the first digital transmissions took place in the 1800s (by means of the wire telegraph), engineers have invented numerous forms, or *modes*, of binary communication. Three classic examples follow.

1. *Morse code* is the oldest two-state means of sending and receiving messages. The logic states are known as *mark* (key-closed or on) and *space* (key-open or off). Morse code is largely obsolete, but some amateur radio operators still use it in their hobby activities.

2. *Baudot*, also called the Murray code, is a five-unit digital code not widely used by today's digital equipment, except in a few antiquated teleprinter systems. There exist 2^5 (that is, 32) possible representations.

3. The *American National Standard Code for Information Interchange* (ASCII) is a seven-unit code for the transmission of text and simple computer programs. There exist 2^7 (that is, 128) possible representations.

Bits, Bytes, and Baud

In a digital computer or communications system, we can represent a bit as either 0 or 1. Some engineers call a group of eight bits an *octet*, and in many systems, an octet corresponds to a unit called a *byte*. We can represent large quantities of data according to powers of 2 or powers of 10. This duality can cause confusion, so beware!

One *kilobit* (kb) equals 1000 (10^3) bits. A *megabit* (Mb) equals 1000 (10^3) kilobits or 1,000,000 (10^6) bits. A *gigabit* (Gb) equals 1000 (10^3) megabits or 1,000,000,000 (10^9) bits. When we express the transfer of data in bits, we always use powers of 10 to define large quantities. If you hear about a modem that operates at 56 kbps, it means 56,000 or 56×10^3 *bits per second* (bps). Bits, kilobits, megabits, and gigabits per second (bps, kbps, Mbps, and Gbps) are commonly used to express the speed of data transmitted from place to place.

Data quantity in storage and memory is usually specified in *kilobytes* (units of 2^{10} or 1024 bytes), *megabytes* (units of 2^{20} or 1,048,576 bytes), and *gigabytes* (units of 2^{30} or 1,073,741,824 bytes). The abbreviations for these units are KB, MB, and GB respectively. Note that the uppercase K represents 2^{10} or 1024, while the lowercase k represents 10^3 or 1000. We always write the prefix-multiplier symbols M and G as uppercase letters, regardless of whether we're working in powers of 2 or powers of 10.

Engineers have introduced progressively larger data units as memory and storage media continue to grow in speed and capacity, as follows:

- One *terabyte* (TB) equals 2^{40} bytes or 1024 GB
- One *petabyte* (PB) equals 2^{50} bytes or 1024 TB
- One *exabyte* (EB) equals 2^{60} bytes or 1024 PB

The term *baud* refers to the number of times per second that a signal changes state. The units of bps and baud are not equivalent, even though people often speak of them as if they meant the same thing.

Still Struggling

These days, we'll hardly ever hear or read about baud (sometimes called the *baud rate*). Engineers and computer scientists prefer to use bits per second (bps) instead. Table 11-1 shows common data speeds and the approximate time periods required to send one page, 10 pages, and 100 pages of double-spaced, typewritten text at each speed.

Forms of Conversion

We can convert analog (continuously variable) data into a string of pulses whose amplitudes have a finite number of states, usually some power of 2. This scheme constitutes *analog-to-digital* (A/D) *conversion*; its reverse is *digital-to-analog* (D/A) *conversion*.

We can transmit and receive binary data one bit at a time along a single line or channel. This mode constitutes *serial data transmission*. We can obtain higher data speeds by using multiple lines or a wideband channel, sending independent sequences of bits along each line or subchannel. In that case, we have *parallel data transmission*.

Parallel-to-serial (P/S) *conversion* involves the reception of bits from multiple lines or channels, and their retransmission one by one along a single line or channel. A *buffer* stores the bits from the parallel lines or channels while they await transmission along the serial line or channel. *Serial-to-parallel* (S/P) *conversion* entails the reception of bits one by one from a serial line or channel, and their retransmission in batches along several lines or channels.

TIP *The output of an S/P converter can't go any faster than the input, but we can find such a system useful when we want to connect or* interface *a serial-data device to a parallel-data device.*

TABLE 11-1. Approximate lengths of time required to send data at various speeds. Abbreviations: s = seconds, ms = milliseconds (units of 0.001 s).

A:

Speed, kbps	Time for One Page	Time for 10 Pages	Time for100 Pages
28.8	0.38 s	3.8 s	38 s
38.4	280 ms	2.8 s	28 s
57.6	190 ms	1.9 s	19 s
100	110 ms	1.1 s	11 s
250	44 ms	440 ms	4.4 s
500	22 ms	220 ms	2.2 s

B:

Speed, Mbps	Time for One Page	Time for 10 Pages	Time for 100 Pages
1.00	11 ms	110 ms	1.1 s
2.50	4.4 ms	44 ms	440 ms
10.0	1.1 ms	11 ms	110 ms
100	0.11 ms	1.1 ms	11 ms

Networks

The *Internet*, often called simply the *Net*, comprises a worldwide system or *network* of computers. It got started in the late 1960s, originally intended to serve as a data storage and communications system that could survive a nuclear war. Its inventors called it the *ARPAnet*, named after the *Advanced Research Project Agency* (ARPA) of the United States federal government.

Protocol and Packets

When people began to connect their computers into ARPAnet, the need became clear for a universal set of standards, called a *protocol*, to ensure that all the machines could "understand" each other. Internet activity consists of computers communicating in a specialized mode known as *machine language*. However, the Internet situation is more complicated than the scenario within a single computer. On the Net, data must usually pass through several different computers to get from the transmitting or *source* computer to the receiving or *destination* computer. These intermediate computers are called *nodes, servers, hosts,* or *Internet service providers* (ISPs).

Millions of people simultaneously use the Net. The most efficient route between a given source and destination can change from moment to moment. Engineers have constructed the Net so that signals always "try" to follow the most efficient possible route. If you're connected to a distant computer, say a machine at the National Hurricane Center, the requests you make of that computer, and the data it sends back to you, are broken into small units called *packets*. Each packet has, in effect, your computer's "name" on it. But not all packets follow the same route through the network. Ultimately, all the packets recombine into the data you want, even though the packets don't necessarily arrive in the same order as the source sent them.

Figure 11-1 is a simplified drawing of Internet data transfer for a hypothetical file containing five packets. Intermediate computers in the circuits, called

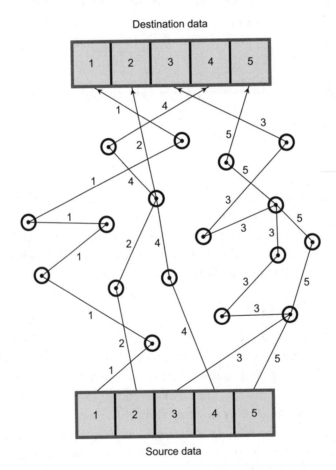

FIGURE 11-1 · Internet data flows in packets from the source to the destination.

nodes, appear as black dots surrounded by circles. The file or program cannot be completely reconstructed until all the packets have arrived.

 **PROBLEM 11-3**

In the communication scenario of Fig. 11-1, suppose that the number of intervening nodes (shown as circles with dots inside) between the source and destination were much greater for each transferred packet. What effect would this additional complexity have?

✔ SOLUTION

It would take somewhat longer for all the packets to reach the destination, and for the complete message to be assembled there. If you were trying to access an informational site on the Net, for example, you would observe an increase in the time between your command and the appearance of the data on your display.

The Modem

The term *modem* is a contraction of *modulator/demodulator*. A modem interfaces a computer to a telephone line, cable network, optical fiber network, or radio transceiver. Figure 11-2 diagrams a modem that allows a computer to

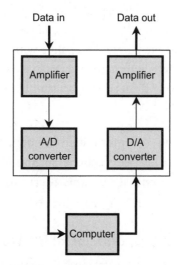

FIGURE 11-2 • Block diagram of a modem for use with a telephone line or cable system.

send and receive data over a telephone line or cable system. The modulator, or D/A converter, changes outgoing digital data into analog waves. The demodulator, or A/D converter, changes incoming analog waves into digital data.

TIP   *When we link multiple computers or other devices in a network, each device has a modem connecting it to the communications medium. The slowest modem determines the maximum speed at which the machines can communicate.*

Getting Information

The Internet can connect you with countless sources of information (and misinformation). When using the Internet for obtaining information, you should know the hour of the day at the remote location, and try to avoid accessing files during peak hours at the remote computer. Peak hours usually correspond to working hours, or approximately 8:00 A.M. to 5:00 P.M. local time, Monday through Friday.

Information-gathering on the Internet takes place on the *World Wide Web* (also called the *Web*), a specialized communications mode within the larger Internet. The outstanding feature of the Web is *hypertext*, a user-friendly scheme for cross-referencing multiple documents. The names of Web sites, known as *uniform resource locators* (URLs), generally begin with the four letters *http*. That abbreviation stands for the technical expression *hypertext transfer protocol*. Certain words, phrases, and images make up *links*.

When you select a link in a *Web page* or *Web site* (a Web-based document containing text, graphics, videos, sounds, and often small programs called *applets*), your computer automatically links to a new site dealing with the same or a related subject. This site will usually contain numerous links of its own. Before long, you might find yourself "surfing the Web" from site to site, simply to satisfy your ever-increasing curiosity!

TIP   *The terms* **Web surfing** *and* **Internet surfing** *derive from the similarity of casual Internet browsing to television* **channel surfing,** *in which the viewer wanders from one channel to another in rapid succession, looking for news or entertainment.*

E-Mail, Newsgroups, Blogs, and Social Networks

For many computer users, communication by means of *electronic mail* (*e-mail*), *newsgroups*, *blogs* (a contraction of *Web logs*), and *social networks* has practically replaced the postal service. You can send messages, pictures, and videos to, and receive them from, friends and relatives scattered throughout the world. To effectively use e-mail, you must have an *e-mail address*. Most newsgroups, blogs, and social media require you to enter your e-mail address as well as your real name or a convenient alias.

At the time of this writing, one of my e-mail addresses is stangibilisco@ hotmail.com. The first part of the address, before the @ symbol, constitutes the *username*. The word after the @ sign and before the last period (or dot) represents the *domain name*. The short abbreviation after the last period indicates the *domain type*. In this case, *com* stands for "commercial." Other common domain suffixes include *net* for "network," *org* for "organization," *edu* for "educational institution," *gov* for "government," and *mil* for "military."

In recent years, country abbreviations have begun to appear at the ends of Internet addresses, such as *us* for "United States," *uk* for "United Kingdom," and *jp* for "Japan." Other abbreviations are constantly coming into common usage, such as *info* for "informational site" and *biz* for "business site."

Internet Conversations

You can carry on a real-time text conversation with other computer users via the Internet. When done among users within a single service provider, this mode is called *chat*. When done among people connected to different service providers, it's called *Internet relay chat* (IRC). When done using small portable devices and numerous abbreviations, it's called *texting*. Typing messages to, and reading them from, other people in real time is more personal than letter writing because your addressees get their messages immediately. But it's less personal than talking on the telephone because you can't hear or make vocal inflections.

Computer users can digitize voice signals and transfer them over the Internet to obtain virtually toll-free long-distance telephone communications. When *Net traffic* is *light* (relatively few people are using a particular portion of the Internet), such connections work well, but when traffic is *heavy* (many people are trying to access the same part of the Internet at the same time), problems occur. All, or nearly all, of the packets must be received and reassembled to get a good signal at the destination. If a significant number of packets arrive disproportionately late, the destination computer will have trouble reassembling an intelligible voice signal.

Local and Wide Area Networks

A *local area network* (LAN) is a group of interconnected computers located in each other's immediate vicinity.

In a *client-server LAN* (Fig. 11-3A), a large computer called a *server* provides a central link among two or more smaller personal computers. In a *peer-to-peer LAN* (Fig. 11-3B), all of the computers have similar computing power, speed, and storage capacity. A peer-to-peer LAN offers greater privacy and user independence than a client-server LAN, but the peer-to-peer scheme is slower when all users share the same data.

A *wide area network* (WAN) is a group of computers linked over a large geographic region. Numerous LANs can be interconnected to form a WAN. The Internet constitutes "the ultimate WAN." Some corporations, universities, and government agencies operate their own proprietary WANs.

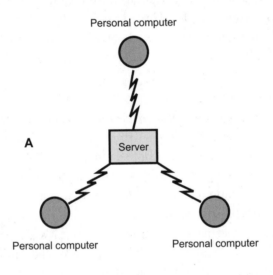

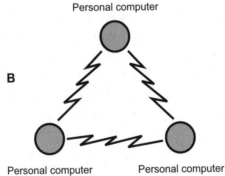

FIGURE 11-3 • At A, a client-server LAN. At B, a peer-to-peer LAN.

Communications Satellites

A *communications satellite* system resembles a gigantic cellular network in which the repeaters orbit the earth, rather than existing at defined locations on the surface. The end users can operate in fixed, mobile, or portable locations.

Geostationary Satellites

At an altitude of approximately 35,800 km (22,200 mi), a satellite in a circular orbit takes one solar day to complete each revolution. If we place a satellite in such an orbit over the equator, and if it revolves in the same direction as the earth rotates, then we have a *geostationary satellite*. From any surface viewpoint, a geostationary satellite stays in the same spot in the sky all the time. A single geostationary satellite can provide coverage of about 40% of the earth's surface. Three such satellites spaced 120° apart in longitude can cover the entire planet except for the extreme polar regions.

Earth-based stations can communicate using a single geostationary satellite only when the stations both lie on direct lines of sight with the satellite. In other words, the applicable satellite must occupy a place in the sky above the horizon as "seen" from both surface locations. If two stations want to communicate and they happen to be nearly on opposite sides of the planet, then they must operate through two geostationary satellites, as shown in Fig. 11-4. That's

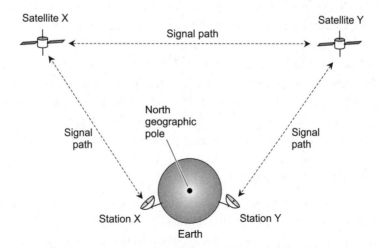

FIGURE 11-4 · A communications link using two geostationary satellites.

because, in such a situation, no geostationary spot exists in space on lines of sight from both locations at the same time.

Still Struggling

Imagine the entire diagram of Fig. 11-4—earth, satellites, and surface-based stations—turning counterclockwise at the rate of one revolution every 24 hours, with the sun in a fixed spot roughly 300 meters (1000 feet) off the top of the page. That mental exercise will give you an idea of what actually takes place as the satellites orbit the earth and the earth rotates "beneath" them.

Low-Earth-Orbit (LEO) Satellites

A satellite in a geostationary orbit requires constant positional adjustment. Geostationary satellites are expensive to launch and maintain. A signal delay always occurs because of the path length. It takes high transmitter power and a precisely aimed dish antenna to communicate reliably. These problems with geostationary satellites have given rise to the *low-earth-orbit* (LEO) satellite concept.

In a LEO system, there are dozens of satellites spaced strategically around the globe in orbits a few hundred kilometers above the surface. The orbits take the satellites near or over the geographic poles. Aerospace engineers call such an orbital path, inclined at nearly 90° to the equator, a *polar orbit*. The satellites relay messages among each other, and to and from end users. If the system has enough satellites, any two end users can maintain constant contact.

A LEO satellite link is easier to use than a geostationary-satellite link. A simple, nondirectional antenna will suffice, and the end user does not have to aim it (or hire a technician to aim it). The transmitter can reach the network using only a few watts of power. Signal-propagation delay, called *latency*, is much shorter with a LEO system than it is with a geostationary system.

Global Positioning System (GPS)

The *Global Positioning System* (GPS) is a network of satellites that allows users to determine their latitude, longitude and (if applicable) altitude above the earth's surface or above sea level. The satellites transmit signals with extremely short wavelengths, comparable to the signals of a police radar. The signals are modulated with timing and identification codes.

A GPS receiver contains a computer that calculates the line-of-sight distances to four different satellites simultaneously. The computer carries out this complicated task by comparing timing codes from signals arriving from each satellite. The computer can thereby provide each user with location data accurate to within a few meters of his or her actual position, based on the four distances.

 PROBLEM 11-4

The average altitude of any satellite tends to change over time. Usually, the orbit decays (the satellite loses altitude). What will occur if a geostationary satellite undergoes this degenerative process, and no one ever bothers to readjust its altitude?

SOLUTION

The orbital period of any satellite gets shorter as the altitude decreases. If a geostationary satellite "falls" to an orbit slightly lower than the prescribed altitude, it speeds up, so it begins to "drift" slowly from west to east as viewed from any fixed point on the earth's surface. The orbital period of any satellite gets longer as the altitude increases. If a geostationary satellite "rises" to an orbit slightly higher than the prescribed altitude, it begins to "drift" slowly from east to west as viewed from any fixed point on the earth's surface.

Personal Communications Systems

Personal communications systems (PCS) include cell phones, pagers, beepers, and all sorts of related paraphernalia. It seems that every day some company comes out with a new device intended for private communications.

Cellular Telecommunications

A *cellular telecommunications* system contains a network of *repeaters*, also known as *base stations*, allowing portable or mobile radio transceivers to operate as telephone sets. A *cell* represents the coverage zone of a single base station. If a cell phone set occupies a fixed location, such as a residence, then communication to and from that phone set usually takes place through a single cell. If the cell phone set is used in a moving vehicle, such as a car or boat, it "wanders" from cell to cell in the network, as shown in the hypothetical example of

Fig. 11-5. In this drawing, the small black dots represent the cellular repeaters, the hexagons represent the cells, and the heavy dashed curve represents the path of the vehicle. All the base stations interface with the telephone system through wires, wireless links, or fiberoptic cables.

 PROBLEM 11-5

Refer to Fig. 11-5. Suppose that you are driving along a roadway, your vehicle's engine fails, and you must pull off the road and stop at a point that happens to lie at the fringes of two adjacent cells. Such a point would show up in Fig. 11-5 as the boundary between two hexagons. You place a call for roadside assistance. The connection occurs, but the signal has an annoying "flutter," making it difficult to carry on a conversation. Why does this "flutter" occur?

SOLUTION

The cellular system "tries" to complete the connection using both of the base stations for which you are at the fringes. The repeaters, therefore, compete with each other. Your signal goes through one repeater and then the other, and then the one again, over and over. This phenomenon is quite common in cellular networks, and it nearly always produces an "echo" or "fluttering" effect technically known as *breaking up*. In recent years, a technology known as *code-division multiple access* (CDMA) has mitigated the severity of situations like this when they occur.

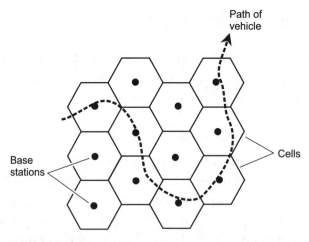

FIGURE 11-5 · In an ideal cellular system, a moving cell-phone set (dashed line) always remains within range of at least one base station.

Wireless Local Loop

A *wireless local loop* (WLL) resembles a cellular system. Telephone sets and data terminals are linked into the system by means of radio transceivers, as shown in Fig. 11-6. Heavy lines represent wire connections, and thin lines represent wireless links. The small ellipses represent subscribers. Each subscriber can use a telephone set or a personal computer equipped with a modem.

Pagers

A simple *pager* or *beeper* employs a small battery-powered radio receiver. Transmitters are located in various places throughout a city, county, or telephone area code district. The receiver picks up a signal that causes the unit to display a series of numerals representing the sender's telephone number.

A pager system equipped with *voice mail* allows the sender to leave a brief spoken message following the beep. A pager receiver equipped to receive e-mail resembles a handheld computer or small calculator. When the unit emits a

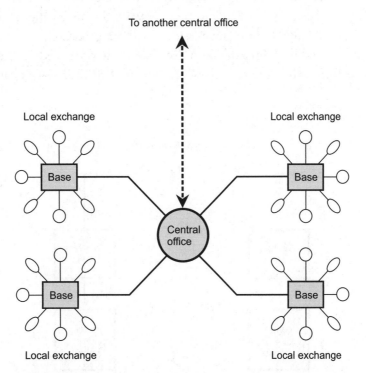

FIGURE 11-6 · Wireless local loop (WLL) telephone systems. Heavy lines represent wire links, and thin lines represent wireless links. Small ellipses represent telephone sets and computers.

sequence of beeps, the user looks at the screen to read text messages. Messages can be stored for later retrieval or transfer to a laptop or desktop computer.

Some pagers can send e-mail messages as well as receive them, using a system similar to a cellular telephone network. Two notes of caution:

1. The use of wireless texting devices may be forbidden on aircraft in flight.

2. Texting should never be attempted while driving a motor vehicle.

Wireless Fax

Facsimile (fax) is a method of sending and receiving nonmoving images. Fax can be done over wireless networks, as well as over the telephone. Figure 11-7 portrays the transmission of a wireless fax signal. Only the sending part of the fax machine is shown at the source (left), and only the receiving part is shown at the destination (right). A complete fax installation has a transmitter and a receiver, allowing for two-way exchange of images.

To send a fax, you place a page of printed material in an *optical scanner*. This device converts the image into binary digital signals (1 and 0). The output of the scanner goes to a modem that converts the binary digital data into an analog signal suitable for transmission. At the destination, the analog signals are converted back into digital pulses like those produced by the optical scanner at the source. These pulses proceed to a printer, data terminal, or computer.

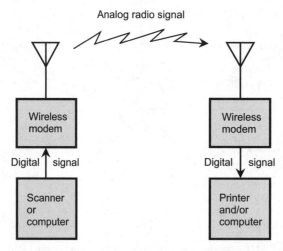

FIGURE 11-7 · Transmission of a wireless fax signal.

Hobby Communications

Hobby communications includes *shortwave listening* (SWLing), the *Citizens' Radio Service*, and *amateur radio*.

Shortwave Listening

A *high-frequency* (HF) radio communications receiver is sometimes called a *shortwave receiver*. Most of these radios function at all frequencies from 3 MHz through 30 MHz. Some also work in the standard AM broadcast band at 535 kHz to 1.605 MHz, and in the spectrum between the AM broadcast band and 3 MHz.

Technically, the shortwave or HF band extends from 3 MHz to 30 MHz. The range of frequencies from 300 kHz to 3 MHz is called the *medium-wave* or *medium-frequency* (MF) band, and the range from 30 kHz to 300 kHz is called the *longwave* or *low-frequency* (LF) band. We can call a receiver that "hears" from 30 kHz to 30 MHz a *general-coverage receiver*.

Some receivers can operate in the ultra-high-frequency (UHF) range from 300 MHz to 3 GHz, as well as at lower frequencies all the way down to a few kilohertz. These high-end radios are called *all-wave receivers*.

TIP   *In the United States, a shortwave listener (or SWLer) need not obtain a license to receive signals. But in some countries, people must obtain a license, even if they want to do nothing more than listen to shortwave broadcasts!*

Citizens' Radio Service

The Citizens' Radio Service, also known as *Citizens' Band* (CB), is a public radio communications and control service. The most familiar form goes by the name *Class D*, which operates on 40 discrete channels near 27 MHz (11 meters) in the HF band.

A 40-channel, 12-W transceiver constitutes the basic radio for class-D fixed-station operation. It employs the SSB voice mode and operates from the standard electric utility. Mobile transceivers also run 12 W SSB, and operate from 13.5-V DC vehicle batteries. The power connection in a mobile installation should run directly to the battery, using a heavy-gauge two-wire cord with an appropriate fuse in series with the positive line. (Most commercially made radios are sold with such cords.) "Cigarette-lighter adapters" can sometimes work for a power connection, but it's best to avoid them because their use increases the risk that the radio transmitter will interfere with the operation of the vehicle's microcomputer.

The organization of *Radio Emergency Associated Communications Teams*, usually called by its acronym, *REACT*, is a worldwide group of radio communications operators. They provide assistance to authorities in disaster areas. On the Class-D band, the emergency channel at 27.065 MHz (channel 9) is monitored by REACT operators.

The *General Mobile Radio Service*, or GMRS, operates at frequencies between 460 and 470 MHz using *Class A* Citizens' Band. The maximum communications range between two individual transceivers in Class A is about 60 km (37 mi). Communications beyond 60 km usually requires the use of repeaters.

TIP *In the United States, some classes of CB operation require government licenses, while others do not. For the latest regulations, consult the manager at an electronics store that sells CB equipment.*

Amateur Radio

A fixed-location amateur-radio (or "ham-radio") station has several components. Figure 11-8 shows a typical example. A computer can facilitate communications by means of *packet radio* with other radio amateurs who own computers. The

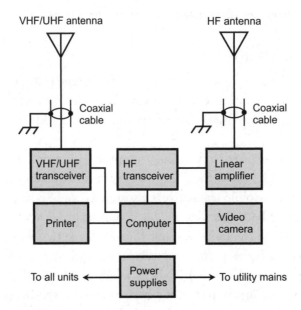

FIGURE 11-8 · A basic amateur-radio station.

station can be equipped for online telephone (landline), cable, wireless, or satellite Internet services. The computer can control the antennas for the station, and can keep a log of all stations that have been contacted. Most modern transceivers can be operated by computer, either locally or by remote control.

Mobile amateur-radio equipment operates in a moving vehicle, such as a car, truck, train, boat, or airplane. Mobile equipment is generally more compact and lightweight than fixed-station apparatus. In addition, mobile gear is designed to withstand large changes in temperature and humidity, as well as mechanical vibration.

Portable amateur-radio equipment almost always operates from battery power and can be set up and dismantled quickly. Some portable equipment can function during physical transport, after the fashion of a cell-phone handset; the *handy-talkie* (HT) or *walkie-talkie* provides a good example. Portable equipment must withstand vibration, temperature and humidity extremes, and prolonged use.

TIP *All amateur radio operation requires licensing.* **The American Radio Relay League (ARRL)** *is the most recognized organization of ham radio operators in the world. They will assist anyone who wants to get an amateur-radio license. You can reach them on the Web at* **www.arrl.org.**

PROBLEM 11-6
One of the amateur-radio bands covers a frequency range of 3500 kHz to 4000 kHz. In what part of the radio spectrum does this span of frequencies lie?

SOLUTION
This band occupies part of the HF range, which covers frequencies from 3 MHz to 30 MHz (3000 kHz to 30 MHz).

Lightning

Lightning constitutes a physical hazard to radio amateurs, CB operators, and shortwave listeners. An outdoor antenna can accumulate a large electrostatic charge during a thundershower. In case of a nearby strike, an *electromagnetic pulse* (EMP) can produce a massive surge of current in an antenna. A direct lightning hit on an antenna produces a catastrophic current and voltage surge that can start fires and electrocute people. Lightning can also induce dangerous voltage "spikes" on utility mains and telephone lines.

The Nature of Lightning

A lightning "bolt," technically called a *stroke*, lasts for a small fraction of a second. The extreme current and voltage liberates a fantastic amount of power for that short time. Four types of lightning exist, as follows:

1. Lightning that occurs within a single cloud (*intracloud*), shown at A in Fig. 11-9

2. Lightning in which electrons flow from a cloud to the earth's surface (*cloud-to-ground*), shown at B in Fig. 11-9

3. Lightning that occurs between two clouds (*intercloud*), shown at C in Fig. 11-9

4. Lightning in which the electrons flow from the earth's surface to a cloud (*ground-to-cloud*), shown at D in Fig. 11-9

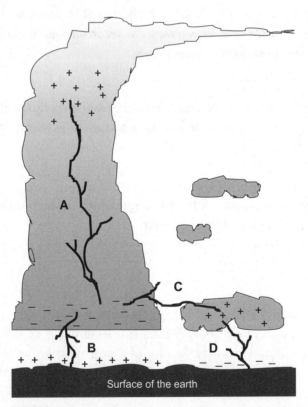

FIGURE 11-9 • Four types of lightning stroke: intracloud (A), cloud-to-ground (B), intercloud (C), and ground-to-cloud (D).

Cloud-to-ground and ground-to-cloud lightning present the greatest danger to electronics hobbyists and equipment. Intracloud or intercloud lightning can cause an EMP sufficient to damage sensitive apparatus, particularly if the antenna has long elements connected directly to a radio. Some radio amateurs and SWLers use *longwire antennas* that can accumulate a substantial electrostatic charge during unstable weather, and that can pick up the EMP from a lightning stroke several kilometers distant.

Protecting Yourself

You can take the following precautions to minimize the hazard to yourself in and near thundershowers. These measures will not guarantee immunity, however. As an old saying goes, "Lightning has a mind of its own." Sometimes, lightning defies logic and seems to operate outside the laws of physics—so beware!

- Remain indoors, or inside a metal enclosure, such as a car, bus, or train.
- Stay away from windows, open or closed, with coverings or without.
- If you can't get indoors, find a low-lying spot on the ground, such as a ditch or ravine, and squat down with your feet close together and your head between your legs until the threat has passed.
- Avoid lone trees or other isolated, tall objects, such as utility poles or flagpoles.
- Avoid electric appliances or electronic equipment that makes use of the utility power lines, or that has an outdoor antenna.
- Stay out of the shower or bathtub.
- Avoid swimming pools, either indoors or outdoors.
- Don't use hard-wired telephone sets.
- Don't use computers with external modems connected, or that operate from the AC utility lines.

Protecting Hardware

Precautions that minimize the risk of damage to electronic equipment (but can't guarantee absolute immunity), particularly radios with outdoor antennas, include the following.

- Never operate an amateur, shortwave, or CB radio station when a thundershower exists near your location.

- Disconnect all antennas and ground all feed line conductors to a good electrical ground other than the utility power-line ground. Leave the lines outside the building and connect them to an earth ground several meters away from the building.

- When you're not using the equipment, unplug it from the utility outlet.

- When you're not using the equipment, disconnect and ground all antenna rotator cables and other wiring that leads outdoors.

- *Lightning arrestors* provide some protection from electrostatic-charge buildup, but they cannot offer complete safety, and should not be relied upon for routine protection.

- *Lightning rods* reduce (but don't eliminate) the chance of a direct hit, but they should not be used as an excuse to neglect the other precautions.

- Power line *transient suppressors* (also called "surge protectors") reduce computer "glitches" and can sometimes protect sensitive components in a power supply, but they should not be used as an excuse to neglect the other precautions.

- Connect antenna supporting masts or towers to an earth ground, using heavy-gauge wire or braid.

- You'll find other secondary protection devices advertised in electronics-related and radio-related magazines.

TIP   *For detailed information about protecting your home appliances against the effects of lightning, consult a competent communications engineer. If you are in doubt about the fire safety of an electronic installation, consult your local fire inspector.*

Still Struggling

Some people express surprise when they learn that a lightning strike on a power line can damage electronic equipment that's plugged into wall outlets, even when the equipment is switched off. The induced surge, which can attain peak values of several thousand volts, arcs (jumps) across the power-switch contacts. When this voltage appears across the power-supply transformer primary, it causes a current surge. This surge, in turn, produces a high-voltage "spike" across the transformer secondary. This "spike" is transmitted to the internal components of the system.

Security and Privacy

People have legitimate concerns about the security and privacy of information exchanged over telecommunications systems because unauthorized persons can obtain and misuse such information.

Wireless versus Wired

Wireless eavesdropping differs from *wiretapping* in two fundamental ways. First, eavesdropping is easier to do in wireless systems than in hard-wired systems. Second, eavesdropping of a wireless link can be impossible to detect, but a competent engineer can usually find and locate a tap in a hard-wired system.

If any portion of a communications connection occurs over a wireless link, then an eavesdropping receiver can be positioned within range of the RF transmitting antenna and the signals intercepted. The existence of the *wireless tap* has no effect on the behavior of equipment in the system. Figure 11-10 shows an example of eavesdropping on a cellular telephone connection.

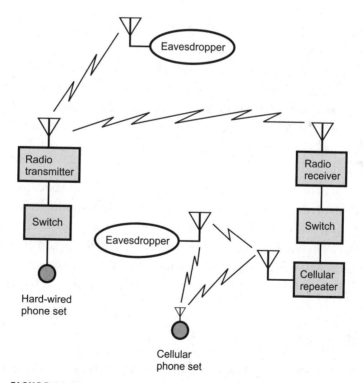

FIGURE 11-10 · Eavesdropping on RF links in a telephone system. Heavy, straight lines represent wires or cables; zig-zags represent RF signals.

Levels of Security

Engineers recognize four levels of telecommunications security, ranging from level 0 (no security) to the most secure connections that technology allows.

1. No security (level 0)—In a *level-0-secure communications system,* anyone can eavesdrop on a connection, assuming that they have the money and/or time to obtain the necessary equipment. Examples of level-0 links include amateur-radio and Citizens' Band (CB) voice communications. The lack of privacy is compounded by the fact that if someone eavesdrops, none of the communicating parties can detect the intrusion under normal circumstances.

2. Wire-equivalent security (level 1)—A *level-1-secure communications system* must be affordable, and it must remain reasonably safe for transactions such as credit-card purchases. When network usage is heavy, the degree of privacy afforded to each subscriber should not decrease, relative to the case when network usage is light. All ciphers should remain unbreakable for at least 12 months, and preferably for 24 months or more. Encryption technology should be updated at least every 12 months, and preferably every six months.

3. Commercial-grade security (level 2)—Some financial and business data warrant protection beyond wire-equivalent. Many companies and individuals refuse to transfer money by electronic means because they fear that criminals will gain access to an account. In a *level-2-secure communications system,* the encryption used in commercial transactions should be such that it would take a hacker at least 10 years, and preferably 20 years or more, to break the cipher. Encryption should be updated as often as economics allow.

4. Military-grade security (level 3)—Security to military specifications (mil spec) involves the most sophisticated encryption available. Technologically advanced countries and entities with economic power have an advantage. The encryption in a *level-3-secure system* should be such that engineers believe it would take a hacker at least 20 years, and preferably 40 years or more, to break the cipher. Encryption should be updated as often as economics allow.

Extent of Encryption

Security and privacy are obtained by *digital encryption* that renders signals readable only to receivers with the necessary *decryption key.*

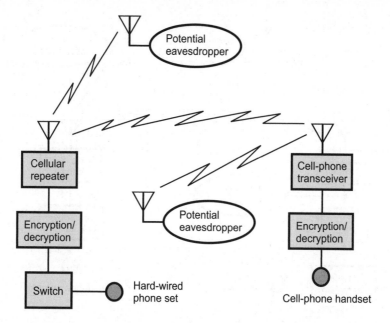

FIGURE 11-11A · Wireless-only encryption. Heavy, straight lines represent wires or cables; zig-zags represent RF signals.

- For level-1 security, encryption is required only for the wireless portion(s) of the circuit. The cipher should be changed at regular intervals to keep it "fresh." The block diagram of Fig. 11-11A shows wireless-only encryption for a hypothetical cellular telephone connection.

- For security at levels 2 and 3, *end-to-end encryption* is necessary. The signal is encrypted at all intermediate points, even those for which signals propagate by wire or cable. Figure 11-11B shows this scheme in place for the same hypothetical cellular connection as depicted in Fig. 11-11A.

Security with Cordless Phones

If someone knows the frequencies (let's call them X and Y) at which a base unit and cordless handset operate, and if that person wants to eavesdrop on conversations that take place over that system, then that person can place a *wireless tap* on the line. The conversation can be intercepted and recorded at a remote site, as shown in Fig. 11-12. It's more difficult to design and construct a wireless tap for a multiple-channel cordless set, as compared with a single-channel set. But if a determined eavesdropper knows all the channel

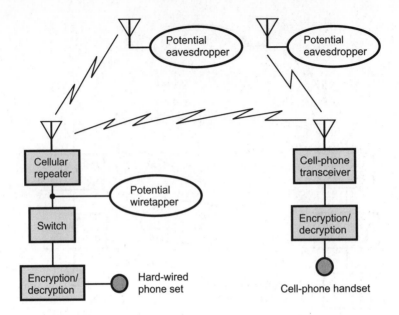

FIGURE 11-11B · End-to-end encryption. Heavy, straight lines represent wires or cables; zig-zags represent RF signals.

frequencies, a *scanning receiver* makes it possible to monitor the frequencies simultaneously and continuously.

TIP   *If you have any security-related worries about using a cordless phone, use a hard-wired phone set instead. They offer an advantage beyond security: They'll usually keep working even when the electric utility fails, while a cordless base unit needs electricity from a "wall outlet" to operate. Of course, hard-wired phones afford less convenience and less flexibility than cordless phones because the cords limit your mobility. You must decide when security considerations override convenience.*

Security with Cell Phones

In effect, cellular telephones act as long-range cordless phone sets. The wider extent of geographical coverage increases the risk of eavesdropping and unauthorized use. Digital encryption of the actual conversation (both ends) offers an effective way to maintain privacy and security of cellular communications. Any technology short of digital encryption leaves a conversation subject to interception.

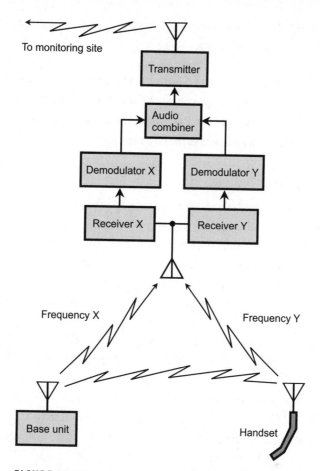

FIGURE 11-12 · Wireless tapping of a cordless telephone.

If we want a cell-phone system to offer the maximum possible level of security, then we must employ access and privacy codes in addition to data encryption. If an unauthorized person knows the code with which a cell phone set accesses the system (the "name" of the set), other sets can be programmed to fool the system into "thinking" the bogus sets belong to the user of the authorized set. Engineers call this activity *cell-phone cloning*.

In addition to digital encryption of data, *user identification* (user ID) must be employed. The simplest such scheme comprises a *personal identification number* (PIN). More sophisticated systems can employ *voice-pattern recognition*, in which the phone set functions only when the designated user's voice enters it.

Hand-print recognition, thumb-print recognition, facial recognition, or *iris-pattern recognition* can also be employed.

TIP *Hand-print, thumb-print, voice-pattern, and iris-pattern recognition constitute examples of* **biometric security technology. The word "biometric" means "measurement of biological characteristics." We should expect to see increasing use of biometric security measures in the coming years.**

Audio Scrambling

Audio scrambling is an older, less sophisticated technology than modern encryption, and has greater vulnerability to interception. If you have an amateur-radio SSB transmitter and a separate receiver capable of operating on the same frequencies as the transmitter, you can demonstrate basic scrambling. Connect the transmitter to a *dummy antenna*, a device designed to dissipate but not radiate or intercept signal power. Set the transmitter for USB operation. Set the receiver for LSB, and tune it to a frequency 3.0 kHz higher than the transmitter. Connect the receiver to another dummy antenna and place it near the transmitter. Figure 11-13A shows the device interconnections. Figure 11-13B shows how the method works.

Suppose that the suppressed-carrier frequency of the USB transmitter equals 3.8000 MHz, producing a spectral output similar to that shown in the graph on the left. Each horizontal division represents 500 Hz, and each vertical division represents 5 dB. You tune the receiver to 3.8030 MHz. The signal energy from the transmitter falls into the receiver passband, as shown by the graph on the right, but the AF components come out "upside-down" within the voice passband. The signal will sound like something from another star system! Engineers call it "monkey chatter." You won't understand any of it. If you want to demonstrate unscrambling, you can record the "monkey chatter" as an audio file. Then you can play the recorded audio back into the transmitter's microphone input. The receiver will produce an AF inversion of an AF inversion. The signal will come out "right-side-up," and you'll hear the original speech.

TIP *Sophisticated audio scramblers split up the audio passband into two or more sub-passbands, inverting some or all of the segments, and rearranging them frequency-wise within the main passband. But audio scrambling constitutes an analog mode, and it can't provide the levels of protection available with digital encryption.*

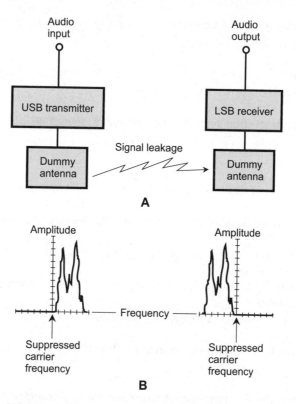

FIGURE 11-13 · A simple voice scrambling and descrambling circuit (at A) and spectral illustrations of the signals (at B).

 Still Struggling

You might ask, "What's the difference between encoding and encryption?" We can define encoding as the translation of data into an orderly form (a code) according to a widely known protocol. The oldest example is the Morse code, in which each character comprises a unique set of short and long pulses called *dots* and *dashes*. No attempt is made to conceal the information from anyone. In contrast, engineers define encryption as the manipulation of data using a cipher, with the intent to conceal the content from everyone except authorized persons or entities. In order to retrieve encrypted data, the receiving machine or operator must have, and properly deploy, a *decryption key* that usually constitutes a specialized computer program.

QUIZ

Refer to the text in this chapter if necessary. A good score is eight correct. The answers are listed in the back of the book.

1. In the octal numeration system, what number follows 999?
    A. 1000
    B. 99A
    C. A000
    D. We can't answer this question because 999 doesn't represent an octal number.

2. What's the principal qualitative difference between a megabyte (MB) and a megabit per second (Mbps)?
    A. We use the former unit to quantify data storage and the latter unit to quantify data transfer speed.
    B. We express the former unit as a binary number and the second unit as a hexa-decimal number.
    C. We use the former unit to quantify serial data and the latter unit to quantify parallel data.
    D. We use the former unit when working with analog data and the latter unit when working with digital data.

3. Which of the following processes *does not* represent a biometric security method?
    A. Iris-pattern recognition
    B. Username recognition
    C. Hand-print recognition
    D. Voice recognition

4. Engineers can (but rarely do) express the rapidity with which a digital signal changes state in term of
    A. kilobits.
    B. baud.
    C. megabytes.
    D. kilobyte-seconds.

5. Which of the following characteristics represents a significant advantage of a LEO-satellite network over a geostationary-satellite network in day-to-day use?
    A. Reduced latency
    B. Greater data transfer rate
    C. Superior reliability
    D. All of the above

6. Suppose that a circuit receives digital data bits one at a time from a single channel, processes the bits, and retransmits them in batches along eight lines or channels. This circuit constitutes

    A. a parallel-to-serial converter.

    B. a digital-to-analog converter.

    C. a serial-to-parallel converter.

    D. an analog-to-digital converter.

7. Which of the following communications systems would we expect to have the greatest level of security?

    A. Amateur radio

    B. Commercial wireless

    C. Cellular telephone

    D. Military wireless

8. The decimal quantity 64 equals the octal quantity

    A. 64.

    B. 32.

    C. 77.

    D. 100.

9. The characters following the last dot in an e-mail address tell us the

    A. username.

    B. domain name.

    C. domain type.

    D. data transfer method.

10. The decimal quantity 129 equals the binary quantity

    A. 1111111.

    B. 11111111.

    C. 1000001.

    D. 10000001.

chapter 12

Communications Antennas

All RF wireless devices need antennas to transmit or receive signals. In this chapter we'll learn how antennas work, and examine a few common types.

CHAPTER OBJECTIVES

In this chapter you will

- See how radiation resistance affects antenna behavior.
- Compare open-dipole, folded-dipole, and zepp antenna designs.
- See how a quarter-wavelength vertical element can serve as an antenna.
- Learn the importance of effective ground systems.
- Discover how phased and parasitic elements affect antenna directivity.
- Compare antenna designs for use in microwave communications.
- Learn how standing waves can affect antenna system performance.

Radiation Resistance

Imagine that you substitute a plain, reactance-free resistor for a transmitting antenna that doesn't have any capacitance or inductance. Suppose that the radio transmitter behaves in the same manner when connected to the resistor as it does when connected to the antenna. For any antenna operating at a certain frequency, there exists a specific resistance, in ohms, for which you can carry out this action. Engineers call it the *radiation resistance* (R_R) of the antenna at the frequency in question.

Determining Factors

Imagine that you place a thin, straight vertical wire over a gigantic sheet of metal that forms a perfectly conducting ground. You connect the antenna output terminals of a radio transmitter between the sheet of metal and the bottom end of the wire. The value of R_R depends on the wire height in wavelengths. Figure 12-1A shows a graph of the function.

Now suppose that you place a length of thin, straight, lossless wire in empty space, cut the wire in the exact center, and connect the antenna output terminals of a radio transmitter at that point. In this case, R_R is a function of the overall conductor length in wavelengths (Fig. 12-1B). As we learned in Chap. 9, the wavelength of a wireless signal depends on the frequency. If f_{MHz} represents

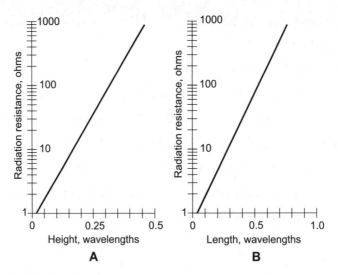

FIGURE 12-1 · Approximate values of radiation resistance for vertical antennas over perfectly conducting ground (A) and for center-fed antennas in free space (B).

the frequency in megahertz and λ_m represents the free-space EM wavelength in meters, then

$$\lambda_m = 300/f_{MHz}$$

and

$$f_{MHz} = 300/\lambda_m$$

If λ_{ft} represents the wavelength in feet, then

$$\lambda_{ft} = 984/f_{MHz}$$

and

$$f_{MHz} = 984/\lambda_{ft}$$

Antenna Efficiency

Ideally, a radio transmitting antenna should have large R_R because radiation resistance always appears in series with *loss resistance* (R_L). An earth ground never conducts current perfectly, and neither do the wires or other elements in a practical antenna. These constraints limit an antenna's *efficiency*, a figure that quantifies its ability to do its job. In mathematical terms, we can express the efficiency (Eff) of an antenna as

$$Eff = R_R/(R_R + R_L)$$

which tells us the ratio of the radiation resistance to the total antenna system resistance. As a percentage, we can calculate the antenna efficiency $(Eff_\%)$ as

$$Eff_\% = 100 \, R_R/(R_R + R_L)$$

PROBLEM 12-1

Suppose that a vertical antenna has a radiation resistance of 37 ohms, but the ground and surrounding objects produce a loss resistance of 111 ohms. (In regions with dry or sandy soil, the loss resistance can easily exceed 100 ohms.) What's the antenna efficiency in percent? What does this figure tell us, in terms of transmitted RF signal power lost?

SOLUTION

In this case, $R_R = 37$ ohms and $R_L = 111$ ohms. When we plug these numbers into the formula for antenna efficiency in percent, we get

$$Eff_\% = 100 \times 37/(37 + 111)$$
$$= 100 \times 37/148$$
$$= 100 \times 0.25$$
$$= 25\%$$

Only 25% of the RF power or energy supplied to the antenna gets radiated into space so that distant receivers can eventually "hear" it! The other 75% of the RF signal power dissipates as heat in the loss resistance.

Half-Wave Antennas

Antennas measuring a half wavelength ($\lambda/2$) from end to end perform well when properly designed and located. Let's look at some basic half-wave antenna characteristics.

Formulas

We can calculate the span of a half wavelength in free space using the formula

$$L_m = 150/f_{MHz}$$

where L_m represents the linear distance in meters, and f_{MHz} represents the frequency in megahertz. If we want to calculate the distance L_{ft} in feet rather than meters, we can use the formula

$$L_{ft} = 492/f_{MHz}$$

TIP   *For ordinary wire, we must multiply the foregoing results by a velocity factor of 0.95 (95%). Engineers use the letter v to represent the velocity factor in equations and formulas. For metal tubing, v can range down to approximately 0.90 (90%). The need for this correction lies in the fact that RF waves travel only 90% to 95% as fast along an antenna element as they do through empty space.*

Open Dipole

An *open dipole* or *doublet* consists of a half-wavelength radiator, such as a straight span of wire or metal tubing, fed at the center, as shown in Fig. 12-2A. Each side or "leg" of the antenna, therefore, measures 1/4 wavelength ($\lambda/4$) long. For a straight wire, we can calculate the approximate length L_m in meters at a design frequency f_{MHz} in megahertz using the formula

$$L_m = 143/f_{MHz}$$

The length in feet is roughly

$$L_{ft} = 467/f_{MHz}$$

These formulas assume a velocity factor of $v = 0.95$ (95%). An open dipole operates effectively at all odd-numbered harmonics of the fundamental frequency, as well as at the fundamental frequency itself.

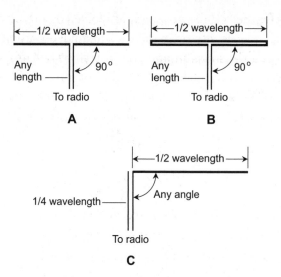

FIGURE 12-2 · Basic half-wave antennas. At A, the open dipole. At B, the folded dipole. At C, the zepp.

PROBLEM 12-2

What's the end-to-end length of a half-wavelength open dipole constructed of wire, intended for use at 7.100 MHz? Express the answer in meters, and also in feet.

SOLUTION

To calculate the end-to-end length of the antenna, we input the numbers to the formulas, getting

$$L_m = 143/7.100$$
$$= 20.1 \text{ m}$$

and

$$L_{ft} = 467/7.100$$
$$= 65.8 \text{ ft}$$

Folded Dipole

A *folded dipole* comprises a half-wavelength, center-fed, radiating element constructed from two parallel wires with their ends connected together as shown in Fig. 12-2B. The folded dipole works well in applications where gain and directivity don't matter much. Perhaps you've seen this type of antenna sold for use as an indoor receiving antenna for commercial FM broadcasts.

The formulas for the end-to-end length of a folded dipole are almost the same as those for a half-wavelength open dipole. The folded portion of the antenna doesn't operate as a transmission line, but instead, it radiates and receives RF signals. You'll need to go through a "trial-and-error" process if you want to build and optimize a folded-dipole antenna. The following formulas represent good approximations, based on a velocity factor of 0.925 (92.5%):

$$L_m = 139/f_{MHz}$$

and

$$L_{ft} = 455/f_{MHz}$$

A folded dipole, like an open dipole, works well at all odd-numbered harmonics of the fundamental frequency, in addition to the fundamental frequency itself.

Zepp

A *zeppelin antenna*, also called a *zepp*, consists of a half-wavelength radiator fed at one end with a quarter-wavelength section of two-wire transmission line, as shown in Fig. 12-2C. For a straight wire radiator, we can calculate the approximate length using the same formulas as we do for a half-wavelength open dipole. For a design frequency f_{MHz}, we can calculate the length L_m in meters as about

$$L_m = 143/f_{MHz}$$

and the length L_{ft} in feet as roughly

$$L_{ft} = 467/f_{MHz}$$

A zepp operates efficiently at all harmonics (both the even-numbered and odd-numbered multiples) of the design frequency, as well as at the design frequency itself. The length of a zepp's radiating element is quite critical—more so than in the case of an open or folded dipole. If the radiating element doesn't have an end-to-end span of almost exactly 0.500 electrical wavelength or a whole number multiple thereof, the parallel-wire feed line won't work properly.

Quarter-Wave Antennas

Antennas measuring a quarter wavelength ($\lambda/4$) are common in wireless communications. A quarter-wavelength antenna will function well if properly designed and located. Most communications specialists orient quarter-wave radiating elements vertically.

Formulas

In free space, a quarter wavelength relates to frequency according to the equation

$$H_m = 75.0/f_{MHz}$$

where H_m represents a quarter wavelength in meters, and f_{MHz} represents the frequency in megahertz. If we express the height of the antenna in feet as H_{ft}, then the formula is

$$H_{ft} = 246/f_{MHz}$$

If v represents the velocity factor in a medium other than free space, then the foregoing formulas modify to

$$H_m = 75.0v/f_{MHz}$$

and

$$H_{ft} = 246v/f_{MHz}$$

For a typical wire conductor, $v = 0.95$ (95%); for metal tubing, v can range down to approximately 0.90 (90%). For a vertical antenna constructed from aluminum tubing such that $v = 0.92$, for example, the formulas are

$$H_m = 69.0/f_{MHz}$$

and

$$H_{ft} = 226/f_{MHz}$$

TIP   *A quarter-wave antenna must always operate against a good ground system for RF. We must do everything that we can (within reason) to minimize the ground loss. Otherwise, the overall antenna system efficiency suffers.*

Ground-Mounted Verticals

The simplest possible vertical antenna comprises a quarter-wave radiator mounted at ground level. The radiator is fed with *coaxial cable*. The center conductor of the cable connects to the bottom end (or base) of the radiator, and the shield of the cable connects to a ground system that includes the earth itself.

At low and medium frequencies, full-size quarter-wave verticals are difficult to construct and maintain because of their considerable height. A technique called *inductive loading* can reduce the physical length of the radiator, while not affecting the electrical length.

We can make a quarter-wave vertical antenna function on several frequencies by inserting multiple *loading coils*, or by inserting parallel-resonant *LC* circuits known as *traps*, at specific points along the radiating element. The design of such an antenna poses a challenge because adjusting any of its resonant frequencies changes all the others.

Unless a ground-mounted vertical antenna has an extensive *ground radial* system installed, we should expect it to exhibit poor efficiency. Inductive loading worsens this situation because it involves the use of a physically shortened radiator whose R_R is lower than that of a full-height quarter-wave radiator.

Vertical antennas typically receive more human-made noise than horizontal antennas. In addition, the EM fields from ground-mounted transmitting antennas are more likely to cause interference to nearby consumer electronic devices than are the EM fields from antennas installed at a height of 1/4 wavelength or more above the ground surface.

Ground-Plane Systems

A *ground-plane antenna* is a vertical radiator (typically measuring 1/4 wavelength in height) operated against a system of 1/4-wave radials elevated at least 1/4 wavelength above the earth's surface. When we place the antenna base that high, we need only three or four radials to obtain a low-loss ground system at RF.

Figure 12-3A illustrates a ground-plane antenna in which the radiating element measures 1/4 wavelength and the radials extend outward horizontally from the feed point. A coaxial-cable feed line transfers RF energy to the antenna from the transmitter. The feed line also sends incoming signal energy from the antenna to the receiver. In some ground-plane antennas, the radials "droop" downward at an angle that can range up to 60° or 70° with respect to the horizon.

In an alternative to the ground-plane design, the radials extend straight down from the feed point in the form of a quarter-wave, cylindrical *sleeve* surrounding, and concentric with, the coaxial feed line. The coaxial cable runs up

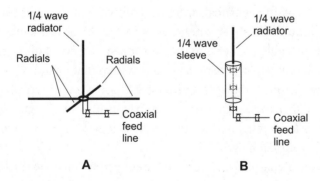

FIGURE 12-3 · Basic quarter-wave vertical antennas. At A, the ground-plane design. At B, the coaxial design.

through the sleeve as shown in Fig. 12-3B. Engineers call this type of system a *coaxial antenna*.

 **PROBLEM 12-3** _____

How tall should we make a quarter-wavelength vertical antenna, constructed of aluminum tubing, for use at a frequency of 50.1 MHz? Express the answer in meters, and also in feet. Assume that the radiating element's velocity factor equals 0.900 (90.0%).

 **SOLUTION** _____

When we input the numbers to the formulas given above and then work through the arithmetic, we obtain

$$H_m = 75v / f_{MHz}$$
$$= 75.0 \times 0.900 / 50.1$$
$$= 1.35 \text{ m}$$

and

$$H_{ft} = 246v / f_{MHz}$$
$$= 246 \times 0.900 / 50.1$$
$$= 4.42 \text{ ft}$$

Loop Antennas

Any receiving or transmitting antenna, consisting of one or more turns of wire forming a DC short circuit at the feed point, constitutes a *loop antenna*. We can categorize these antennas as *small loops* or *large loops*.

Small Loop

A *small loop antenna* has a circumference of less than 0.1 electrical wavelength. It works well for receiving but not for transmitting, unless we take extraordinary measures to minimize the resistance in the loop conductor and associated components. A small loop has the poorest signal response along its axis (the line perpendicular to the plane containing the loop), and the best signal response in the plane containing the loop.

We can connect a capacitor in series or parallel with a small loop to produce resonance at a specific frequency, thereby improving the sensitivity at that frequency. Figure 12-4 shows a variable capacitor in parallel with a small loop, allowing for adjustment of the resonant frequency.

TIP   *Small loops work well for radio direction finding (RDF), and also for reducing interference caused by human-made noise or strong local signals at frequencies below about 20 MHz. Along the loop's axis, the null (direction of minimum response) is sharp and deep. As we rotate a small loop, we observe a pronounced decrease in the received signal strength when the axis of the loop points directly toward a nearby transmitting antenna or source of interference.*

Loopstick

For receiving applications at frequencies up to approximately 20 MHz, we can use a *loopstick antenna*, a variant of the small loop. It comprises a coil of insulated or enameled wire, wound on a solenoidal (rod-shaped) powdered-iron core. A series-connected or parallel-connected capacitor, in conjunction with the coil, forms a tuned circuit to enhance the sensitivity and provide some selectivity. If we want this type of antenna to work, we must avoid the temptation to install it inside an equipment cabinet, particularly if that cabinet contains any metal. Loopsticks offer convenience because they can be located near a receiver—even directly on top of it or on the back panel.

A loopstick has directional characteristics similar to those of the small loop shown in Fig. 12-4. The greatest sensitivity occurs off the sides of the coil, in the plane perpendicular to the axis of the core. A sharp null occurs off the ends,

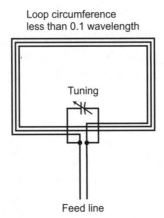

Loop circumference
less than 0.1 wavelength

Tuning

Feed line

FIGURE 12-4 · A small loop antenna with a capacitor for adjusting the resonant frequency.

along a straight line corresponding to the axis of the core. We can use this null to minimize interference from local signals and human-made sources of noise, just as we can do with a small loop. Some commercially manufactured loop-sticks have horizontal (*azimuth*) and vertical (*elevation*) bearings, so that we can orient its null any way we want in three-dimensional space.

Large Loop

If a loop antenna has a circumference greater than 0.1 wavelength, we call it a *large loop*. This type of antenna typically has a circumference of either a half wavelength or a full wavelength, and exhibits resonance all by itself without the need for a tuning capacitor. We can use a large loop for transmitting or receiving.

The maximum radiation and response for a half-wavelength loop occur in the plane containing the loop. We can set up the loop with its axis oriented vertically, so that it radiates and receives fairly well in all horizontal directions.

The maximum radiation and response for a full-wavelength loop occur along the axis. We can install this type of loop antenna with its axis oriented horizontally, so it transmits and receives best in two opposite compass directions. We can then rotate the loop in a horizontal plane to vary its so-called *favored directions* (directions in which it exhibits maximum radiation and response).

Full-Wave Loop Circumference

We can calculate the circumference of a large, square, full-wavelength loop made of wire, with each side measuring 1/4 wavelength long, with the formula

$$C_m = 306/f_{MHz}$$

where C_m represents the circumference in meters, or with the formula

$$C_{ft} = 1005/f_{MHz}$$

where C_{ft} represents the circumference in feet.

Still Struggling

Do you notice something strange about the foregoing formulas? They suggest that, when you bend an antenna into a full-wavelength loop, you should make its physical length longer than an equivalent, straight-line wavelength in free space—as if the velocity factor could somehow exceed 1, meaning that the EM field travels faster than the speed of light! The technical explanation for this anomaly goes beyond the scope of this book. It involves the way the EM field around a full-wavelength loop interacts with itself. You can rest assured that this effect doesn't cause the EM fields to travel around the loop faster than the speed of light, which is the same as the speed of all EM fields in free space (approximately 300,000 km/s or 186,000 mi/s).

 PROBLEM **12-4**

What's the circumference of a full-wavelength loop made of wire and designed for a resonant frequency of 14.0 MHz? Express the answer in meters, and also in feet.

 SOLUTION

Using the foregoing formulas, we can plug in the numbers to get

$$C_m = 306 / f_{MHz}$$
$$= 306 / 14.0$$
$$= 21.9 \text{ m}$$

and

$$C_{ft} = 1005 / f_{MHz}$$
$$= 1005 / 14.0$$
$$= 71.8 \text{ ft}$$

Ground Systems

A well-designed antenna must always have an excellent *DC ground* connection (also called *electrical ground*), along with a low-loss *RF ground* system. These two types of "grounds" are *not* equivalent. Even the best DC earth ground connection can perform poorly for RF purposes, and you can construct an excellent RF ground system without any DC connection to the earth whatsoever.

Electrical Grounding

Electrical grounding ensures the personal safety of people who work on antenna systems. The electrical ground can help to protect equipment from damage if lightning strikes nearby (although it can't *guarantee* that nothing bad will happen). The DC ground also minimizes the risk of serious *electromagnetic interference* (EMI) to and from wireless communications equipment.

A competent engineer always heeds an informal commandment: "Never touch two grounds at the same time." This admonition, while humorous as stated, addresses a serious danger. It refers to the tendency for potential differences to exist between supposedly neutral points in apparent defiance of common-sense electricity and electronics theory. Experienced people have received severe electrical shocks because they forgot this rule. You should always wear electrically insulating gloves and rubber-soled shoes when working with ground systems of any kind.

Some appliances have *three-wire cords* for connection to the electric utility. One wire goes to the "common" or "ground" part of the hardware, and leads to a D-shaped or U-shaped prong in the plug. You should never cut off or defeat this ground prong because such modification can result in dangerous voltages appearing on exposed metal surfaces, and will also void the equipment warranty.

PROBLEM 12-5

What might cause a potential difference to exist between the grounded wire of a three-wire electrical cord and a copper cold-water pipe in your house?

SOLUTION

Unless the copper pipe has a direct, near-zero-resistance connection to the grounded wires in the electrical system, a considerable AC voltage can exist between them. This voltage is induced by *electromagnetic coupling* between the wiring and the plumbing. Electrical currents can also exist in the earth itself, giving rise to potential differences between points in the soil located more than a few meters apart. Even if a cold-water pipe runs directly into the earth and all of the pipe splices are metal-to-metal, an AC voltage can exist between the pipe and the ground wire of the household electrical system. This voltage can reach levels sufficient to cause a lethal electrical shock.

RF Grounding

Figure 12-5 shows a proper RF ground scheme and an improper one. In a good RF ground system (shown at A), each device goes to a common *ground*

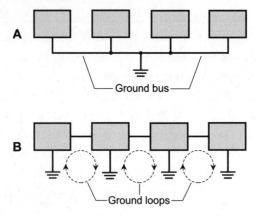

FIGURE 12-5 · At A, the correct method for grounding multiple hardware equipments. At B, an incorrect method for grounding multiple equipments, creating RF ground loops.

bus, which in turn runs to the earth ground by a single conductor of the shortest possible length. A poor ground system (shown at B) contains *ground loops* that increase the susceptibility of equipment to EMI. Long ground wires and ground loops can act as RF antennas, no matter how low we can get the DC resistances.

Ground radials measure 1/4 wavelength or more. They run outward from the base of a surface-mounted vertical antenna, and are connected to the shield of the coaxial feed line. You can lay radials directly on the ground surface or bury them a few centimeters below the surface. As you increase the number of radials having a certain fixed length in a surface-mounted vertical antenna system, the antenna performance improves. Also, as you increase the length of each radial in a system having a certain fixed number of radials, the antenna performance improves.

As stated earlier in this chapter, if you elevate the base of a vertical antenna at least 1/4 wavelength above the ground surface, you only need three or four 1/4-wave radials to obtain a good RF ground.

A *counterpoise* provides an alternative means of obtaining an RF ground or ground plane. You can place a grid of wires, a screen, or a metal sheet above the earth's surface and orient it horizontally, in effect producing a capacitor with the counterpoise as one "plate" and the earth as the other "plate."

TIP   *Ideally, the radius of a counterpoise should measure at least 1/4 wavelength at the lowest frequency that you intend to use.*

Gain and Directivity

We define the *power gain* of a transmitting antenna as the ratio of the maximum *effective radiated power* (ERP) to the actual RF power applied at the feed point, expressed in decibels (dB). If the ERP equals P_{ERP} watts and the applied power equals P watts, then

$$\text{Power Gain (dB)} = 10 \log (P_{ERP}/P)$$

where "log" represents the base-10 logarithm. Engineers specify power gain in the antenna's favored direction(s).

Reference Antennas

In order to define ERP and power gain, we must choose a specific type of *reference antenna* and then define its power gain as 0 dB by default. The ERP for a reference antenna equals the actual applied power. When an antenna has positive power gain in a certain direction, it means that the ERP exceeds the actual applied power in that direction. Negative power gain in a certain direction tells us that the actual applied power exceeds the ERP in that direction.

Whenever we have an antenna with positive power gain in a certain direction, we inevitably find that it has negative power gain (that is, power loss) in some other direction. Antenna gain differs conceptually from amplifier gain! A directional antenna can't *increase* the total amount of power that it gets from the transmitter, no matter how great its power gain figure. A directional antenna can only *redistribute* the overall radiated power, adding it in some directions at the expense of other directions.

A half-wave dipole in free space makes a good reference antenna. Power-gain figures taken with respect to a dipole (in its favored directions) are expressed in units called dBd (decibels with respect to dipole). Alternatively, the reference antenna for power-gain measurements can be an *isotropic radiator*, which radiates and receives equally well in all three-dimensional directions. In this case, units of power gain are called dBi (decibels with respect to isotropic). For any given antenna, the power gains in dBd and dBi always differ by approximately 2.15 dB, as follows:

$$\text{Power Gain (dBi)} = 2.15 + \text{Power Gain (dBd)}$$

Directivity Plots

We can draw diagrams of antenna radiation and response patterns using polar-coordinate plots, such as the ones in Fig. 12-6. We assume that the antenna

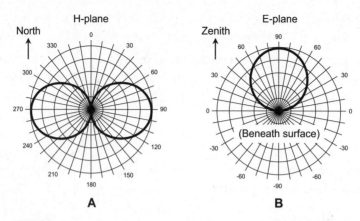

FIGURE 12-6 · Directivity plots for a dipole. At A, the H-plane (horizontal plane) plot as viewed from high above the antenna. Coordinate numbers indicate compass-bearing (azimuth) angles in degrees. At B, the E-plane (elevation plane) plot as viewed from a point on the earth's surface far from the antenna. Coordinate numbers indicate elevation angles in degrees above or below the plane of the earth's surface.

exists at the center, or *origin*, of the coordinate system. We can read the relative gain in any direction by checking to see how far the curve lies from the origin in that direction. The greater the radiation or reception capability of the antenna in a certain direction, the farther from the origin the applicable point on the curve appears. Some engineers use the term *lobe* in reference to a closed loop in the curve of an antenna directivity plot.

A half-wave dipole antenna, oriented horizontally so that its conductor runs in a north-south direction, has a *horizontal-plane* (or *H-plane*) directional pattern similar to the one shown in Fig. 12-6A. In this illustration, the axis of the wire runs along the line connecting 0° and 180°. The angles correspond to directions of the compass, where 0° represents geographic north, 90° represents east, 180° represents south, and 270° represents west. We can see that the greatest gain occurs toward the east and west in this case. Those are the directions to and from which the main (strongest) lobes exist.

The *elevation-plane* (or *E-plane*) pattern of a half-wave dipole depends on how high we put it above the ground surface. With the dipole oriented so that its conductor runs perpendicular to the page and the antenna set up 1/4 wavelength above the ground, the E-plane antenna pattern resembles the graph shown in Fig. 12-6B. The angles represent degrees of elevation relative to the horizon. (The lower half of the coordinate plane represents points below the horizon.) In the scenario of Fig. 12-6B, the greatest gain and response go straight up. The main (and only) lobe lies directly above the antenna.

The patterns shown in Figs. 12-6A and 12-6B are also characteristic of half-wave folded dipole and half-wave zepp antennas oriented horizontally so that the radiating elements run in a north-south direction.

Forward Gain

We define the *forward gain* of an antenna as the ratio, in decibels, of the ERP in its favored direction (main lobe) relative to the ERP from a dipole antenna in its favored direction. Some specialized antennas at very-high and ultra-high frequencies (above about 30 MHz) can exhibit forward gain figures of 20 dBd or more. At microwave frequencies exceeding approximately 3 GHz, large dish antennas can have forward gain figures upwards of 35 dBd. In general, as the frequency rises and the wavelength gets shorter, we find it increasingly practical to construct antennas with exceptional forward gain.

Front-to-Back Ratio

The *front-to-back ratio* of a *unidirectional* (one-way) antenna, abbreviated f/b, gives us a decibel (dB) comparison of the ERP in the center of the main lobe relative to the ERP in the direction precisely opposite the center of the main lobe. Figure 12-7 shows the H-plane directivity plot for a hypothetical unidirectional antenna pointed north. We can deduce the f/b ratio by comparing the ERP between north (0°) and south (180°). If each pair of adjacent, concentric circular graph divisions represents a signal-level difference of 5 dB, then the f/b ratio in this case equals 15 dB (three divisions).

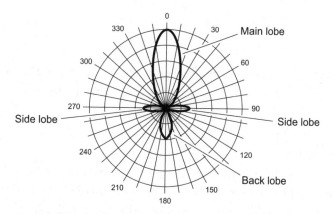

FIGURE 12-7 · Directivity plot for a hypothetical antenna in the H (horizontal) plane. We can determine the front-to-back and front-to-side ratios from such a graph. Coordinate numbers indicate compass-bearing angles in degrees.

Front-to-Side Ratio

The *front-to-side ratio* of an antenna, abbreviated f/s, provides another expression of the directivity of an antenna system. The f/s ratio, like f/b, is expressed in decibels (dB). We compare the ERP in the center of the main lobe with the ERP in the centers of the *side lobes* (the ones at right angles to the favored direction). Let's look again at Fig. 12-7. In this case, we can infer the f/s ratios by comparing the ERP levels between north and east (right-hand f/s), or between north and west (left-hand f/s). They both look like roughly 17.5 dB (three and a half concentric graph divisions).

TIP   *In most antennas operated in free space without any nearby obstructions, the right-hand and left-hand f/s ratios turn out equal. In practice, however, they often differ because electrically conducting objects near the antenna can distort the directional pattern.*

PROBLEM 12-6

Suppose that each pair of adjacent, concentric circles in Fig. 12-7 represents a difference of 10 dB. What's the f/s ratio for the antenna?

SOLUTION

The centers of the side lobes are about three and a half circles inward from the center of the main lobe. Therefore, the f/s ratio in this example equals approximately 35 dB.

Phased Arrays

A *phased array* uses two or more *driven elements* (elements directly connected to the feed line) to produce gain in some directions at the expense of other directions.

End-Fire Array

Figure 12-8 shows two simple pairs of phased dipoles connected together to form so-called *end-fire arrays*. In the system of Fig. 12-8A, we space the dipoles a quarter wavelength apart, orient them parallel to each other, and feed them 90° out of phase, resulting in a unidirectional pattern. We can obtain a *bidirectional* (two-way) pattern by spacing the dipoles a full wavelength apart and feeding them in phase, as shown in Fig. 12-8B.

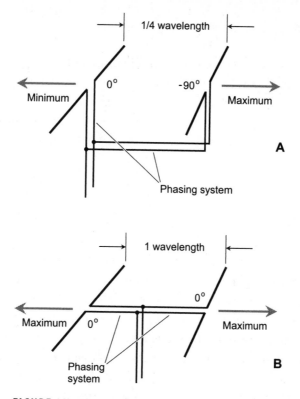

FIGURE 12-8 · At A, a unidirectional end-fire antenna array. At B, a bidirectional end-fire array.

The lengths of the phasing sections are critical to the proper performance of a phased array. In the system shown at Fig. 12-8A, the electrical distance between the junction of the two transmission-line sections and the right-hand antenna should be exactly 1/4 wavelength longer than the distance between the junction and the left-hand antenna. In the system shown at Fig. 12-8B, the electrical distance between the junction of the two transmission-line sections and the right-hand antenna should be the same as the distance between the junction and the left-hand antenna.

We can calculate the physical span of an electrical quarter wavelength using either of the formulas

$$Q_m = 75.0v/f_{MHz}$$

or

$$Q_{ft} = 246v/f_{MHz}$$

where Q_m represents the span in meters, Q_{ft} represents the span in feet, v represents the velocity factor of the transmission line used for the phasing lines (also called the *phasing harness*), and f_{MHz} represents the operating frequency in megahertz.

Phased arrays can have fixed or steerable directional patterns. If the wavelength is short enough to allow construction from metal tubing, we can mount the pair of phased dipoles shown in Fig. 12-8A on a rotator for 360° directional adjustability. With phased arrays of vertical antennas, we can vary the relative signal phase, allowing us to "steer" the directional pattern to some extent, without having to mechanically rotate the system.

Longwire Antenna

Among antenna engineers and hobbyists such as amateur-radio operators, disagreement exists concerning the minimum length that a straight wire must have in order to constitute a true *longwire antenna*. Some people say that one full wavelength is long enough; others insist on two or three full wavelengths. Some folks casually call random-length, end-fed wires "longwires," even if they measure less than a full wavelength from end to end, but such systems are more appropriately called *random-wire antennas*.

In order to qualify as a true longwire, an antenna element must lie entirely along a single straight line (with the exception of some *wire sag* for long open spans), and it must be long enough to produce substantial gain over a dipole. As the wire length increases, so does the gain. A straight longwire measuring six or eight wavelengths from end to end can produce several dBd of gain in its favored directions. The radiation pattern contains multiple *major lobes* and *minor lobes*. The major lobes, where the maximum radiation and response occur, lie nearly in line with the wire. Minor lobes appear in many different directions. As the wire length increases, so does the number of minor lobes.

TIP *If a longwire measures several (or many) wavelengths from end to end, the details of its radiation pattern change dramatically with small variations in wire length at a fixed frequency, or with small variations in frequency for a fixed wire length. However, this theoretical change does not always translate into a noticeable performance difference.*

Broadside Array

Figure 12-9 shows a simple *broadside array*. The driven elements can each comprise a single half-wave dipole radiator, as shown here, or they can be complex systems with individual directive properties. If we place a reflecting screen behind the array of dipoles in the system of Fig. 12-9, we obtain a *billboard antenna* with a unidirectional pattern.

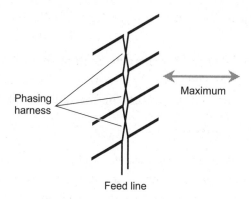

FIGURE 12-9 · A broadside array. The elements all receive their portions of the outgoing signal in phase, and at equal amplitudes.

The directional properties of a broadside array depend on the number of elements, whether or not the elements have gain themselves, the spacing among the elements, and whether or not we include a reflecting screen. In general, as the number of elements increases, so does the power gain in the favored direction(s).

 **PROBLEM 12-7**

What's the length of a quarter-wave transmission line, used as a phasing harness in an antenna similar to the one shown in Fig. 12-8A, if the velocity factor of the line equals 0.800 and we operate the system at 18.1 MHz? Express the answer in meters and in feet.

 **SOLUTION**

We can use the applicable formulas and input our numerical data to get

$$Q_m = 75.0v/f_{MHz}$$
$$= 75.0 \times 0.800/18.1$$
$$= 3.31 \text{ m}$$

where Q_m represents the length in meters, and

$$Q_{ft} = 246v/f_{MHz}$$
$$= 246 \times 0.800/18.1$$
$$= 10.9 \text{ ft}$$

where Q_{ft} represents the length in feet.

Parasitic Arrays

Parasitic arrays are used at frequencies ranging from about 2 MHz up to several gigahertz to get directivity and forward gain with a unidirectional pattern. The two simplest and most common examples are the *Yagi antenna* and the *quad antenna*.

What's a Parasitic Element?

Engineers define a *parasitic element* as an electrical conductor that forms an important part of an antenna system without any DC connection to the feed line. Parasitic elements enhance antenna power gain and directivity by interacting with the EM field from the driven element. We can categorize parasitic elements in two ways, according to the directional effects they produce:

1. When gain occurs in the direction going from the driven element toward the parasitic element, we call the parasitic element a *director*.

2. When gain occurs in the direction going from the parasitic element toward the driven element, we call the parasitic element a *reflector*.

Directors usually (but not always) have end-to-end spans a few percent shorter than the driven element. Reflectors usually measure a few percent longer than the driven element. The driven element itself is usually a half-wave dipole or a full-wave loop.

Yagi Antenna

The *Yagi antenna* is an array of parallel, straight elements. "Yagi" was the name of one of the Japanese inventors of this antenna. The other inventor had the surname "Uda," so you'll sometimes see or hear this type of antenna called a *Yagi-Uda*. We can construct a simple Yagi by placing a director or a reflector parallel to, and a specific distance away from, a single driven element.

The optimum spacing between the elements of a driven-element-and-director Yagi is 0.1 to 0.2 wavelengths, with the director tuned to a frequency 5% to 10% higher than the resonant frequency of the driven element. The optimum spacing between the elements of a driven-element-and-reflector Yagi is 0.15 to 0.2 wavelengths, with the reflector tuned to a frequency 5% to 10% lower than the resonant frequency of the driven element. Either of these designs gives us a *two-element Yagi*. We can expect to get about 5 dBd of forward gain from a well-designed antenna of this type.

A *three-element Yagi* with one director and one reflector, along with the driven element, increases the gain and f/b ratio compared with a two-element

Yagi. An optimally designed three-element Yagi has approximately 7 dBd of forward gain. Figure 12-10 is a conceptual diagram (but not a design blueprint) for a three-element Yagi.

The gain and f/b ratio figures for an optimally designed Yagi increase as we add more elements, usually by placing extra directors in front of a three-element Yagi and keeping only one reflector. Each additional director has an end-to-end span slightly shorter than its predecessor.

TIP   *The design and construction of Yagi antennas having numerous elements poses a significant challenge. As the number of elements increases, the optimum antenna dimensions become more complicated and critical. The performance of any long Yagi antenna should be optimized with field tests before the antenna is placed in service.*

Quad Antenna

A *quad antenna* operates according to the same principles as the Yagi, except that it employs full-wavelength loop elements instead of straight half-wavelength elements.

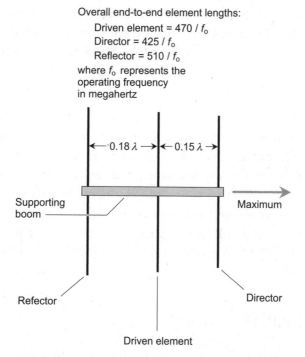

Overall end-to-end element lengths:
  Driven element = 470 / f_o
  Director = 425 / f_o
  Reflector = 510 / f_o
where f_o represents the operating frequency in megahertz

$\leftarrow$ 0.18 λ $\rightarrow$ $\leftarrow$ 0.15 λ $\rightarrow$

Supporting boom

Maximum

Refector

Director

Driven element

FIGURE 12-10 · A three-element Yagi antenna. See text for discussion of specific dimensions.

A *two-element quad* can consist of a driven element and a reflector, or it can have a driven element and a director. A *three-element quad* has one driven element, one director, and one reflector. The director is tuned to a frequency 5% to 10% higher than the resonant frequency of the driven element. The reflector is tuned to a frequency 5% to 10% lower than the resonant frequency of the driven element.

We can add extra directors to the basic three-element quad design to form quads having any desired numbers of elements. The gain increases as the number of elements increases. Each succeeding director is slightly shorter than its predecessor. Alternatively, the additional directors can comprise half-wave, straight elements similar to those of a multielement Yagi. Amateur radio operators have coined the term *quagi* to describe this quad-Yagi hybrid.

TIP *Long quad and quagi antennas are practical only at frequencies above about 100 MHz (a free-space wavelength shorter than roughly 3 m). At longer wavelengths, such antennas become unwieldy because of their large size and mass.*

PROBLEM 12-8

Suppose that we design a two-element Yagi antenna having a driven element and one director. We cut its driven element for a half-wave resonance at a frequency of 10.1 MHz. What's the approximate range of frequencies for which we should trim the director?

SOLUTION

The director in a two-element Yagi should be trimmed for half-wave resonance at a frequency 5% to 10% higher than that of the driven element. Therefore, the half-wave resonant frequency of the director should lie between 1.05 and 1.10 times the resonant frequency of the driven element. The lower limit of this range equals 10.1 × 1.05 = 10.605 MHz, which we can round off to 10.6 MHz. The upper limit of the range equals 10.1 × 1.10 = 11.11 MHz, which we can round off to 11.1 MHz.

Antennas for UHF and Microwaves

At radio frequencies above approximately 300 MHz, high-gain antennas have manageable size and weight. The wavelength is short enough to make antennas of multiwavelength dimensions practical.

Dish Antenna

Everybody knows what a *dish antenna* looks like, but few people appreciate the precision of the design and alignment parameters. The most efficient type of dish employs a *paraboloidal reflector.* That's a section of a *paraboloid* that we get when we rotate a curve called a *parabola* around its axis in three-dimensional space. A somewhat less precise, but in most cases workable, alternative is the *spherical reflector*, whose shape constitutes a section of a three-dimensional sphere.

In *conventional dish feed* (Fig. 12-11A), we place a small receiving preamplifier and transmitting unit at the reflector's focal point. These devices operate by remote control through coaxial cables. Alternatively, the antenna or preamplifier/transmitter can be placed behind the main dish, and the signals reflected from a convex "EM mirror" located at the focal point. Engineers call this design scheme *Cassegrain dish feed* (Fig. 12-11B).

As we increase the diameter of a dish reflector, the forward gain increases and the main lobe grows narrower. At the absolute minimum, a dish antenna

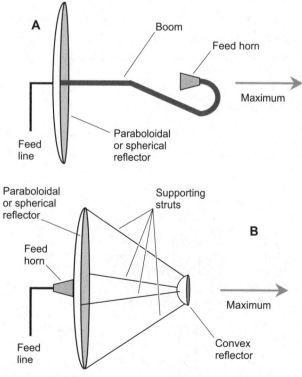

FIGURE 12-11 · Dish antennas with conventional feed (A) and Cassegrain feed (B).

must have a diameter of several wavelengths if we expect to obtain proper operation and reasonable gain. The reflecting element can be made from sheet metal, a screen, or a wire mesh. If a screen or mesh is used, the spacing between adjacent wires must never exceed a small fraction of a wavelength.

Helical Antenna

A *helical antenna* is a high-gain, unidirectional antenna whose shape resembles a spring, often in conjunction with a flat reflecting screen, as shown in Fig. 12-12. A helical antenna can operate over a wide band of frequencies above a specific lower limit.

For a helical antenna to perform at its best, the reflector should measure at least 0.8 wavelength in diameter, which is the lowest operating frequency. The radius of the helix should equal approximately 0.17 wavelength at the center of the intended operating frequency range. The longitudinal spacing between turns should be roughly 0.25 wavelength at the center of the operating frequency range. The entire helix should measure at least 1.0 wavelength from end to end at the lowest operating frequency.

A helical antenna can provide about 15 dBd forward gain. Groups (or *bays*) of helical antennas, all placed side-by-side in a square matrix with a single large reflector, are common in space- and satellite-communications systems. Helical-antenna bays offer greater gain and better directivity than a single helical antenna can produce.

Corner-Reflector Antenna

Figure 12-13 shows an example of a simple *corner-reflector antenna* with a half-wave driven element. This system provides modest gain over a half-wave dipole.

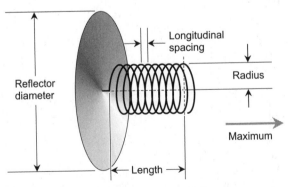

FIGURE 12-12 · A helical antenna with a flat reflector.

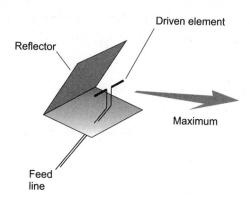

FIGURE 12-13 · A corner reflector that employs a dipole antenna as the driven element.

The reflector can consist of wire mesh, a screen, or a piece of sheet metal. Corner reflectors are used for television (TV) reception in locations where cable is not available and the user does not wish to pay for a satellite TV service. We'll also see antennas of this type employed in microwave links, such as the ones that interconnect highway call boxes with central stations. Several half-wave dipoles can be fed in phase and placed along a common axis with a single, elongated reflector, forming a *collinear corner-reflector array*.

Horn Antenna

Several different configurations exist for the *horn antenna*, but they all look similar. Figure 12-14 is a representative drawing. This antenna provides a unidirectional radiation and response pattern, with the favored direction coincident with the opening of the horn. The preferred feed system comprises a *waveguide* (which we'll examine in a little more detail shortly) that joins the

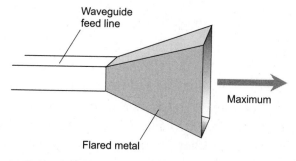

FIGURE 12-14 · A horn antenna, commonly used at microwave frequencies.

antenna at the narrowest point of the horn. Horn antennas perform best at extremely short wavelengths. They find extensive application at microwave frequencies.

TIP   *Horns are sometimes used to feed large dish antennas. The horn points backward toward the center of the dish reflector. This arrangement "sharpens" the main lobe, minimizing extraneous received-signal response and transmitted-signal radiation.*

 PROBLEM 12-9

What's the minimum overall length a helical antenna should have if we want it to work effectively at 900 MHz?

 SOLUTION

The overall helix length should be at least 1.0 wavelength in free space. Recalling the free-space wavelength formulas given earlier in this chapter, and "plugging in" the numbers, we get

$$\lambda_m = 300 / f_{MHz}$$
$$= 300 / 900$$
$$= 0.333 \text{ m}$$

for the minimum helix length in meters, and

$$\lambda_{ft} = 984 / f_{MHz}$$
$$= 984 / 900$$
$$= 1.09 \text{ ft}$$

for the minimum helix length in feet.

Transmission Lines

We can employ an RF *transmission line* to transfer a signal from one place to another, usually from a wireless transmitter to an antenna or from an antenna to a wireless receiver. An ideal transmission line carries the signal, which exists in the form of an EM field, without radiating any of it, and without picking up any external EM energy either. The most common types are *parallel-wire lines, coaxial lines,* and *waveguides.*

Parallel-Wire Line

The "twin lead" or "ribbon cable" that goes from an old-fashioned television (TV) receiving antenna to the TV set is an example of parallel-wire line. As the term suggests, parallel-wire line comprises two identical wires running alongside each other, spaced at a constant distance. It's sometimes called *balanced line* because the two conductors carry equal and opposite RF currents whose EM fields theoretically balance each other, preventing the line from radiating or intercepting signals. (In practice, the balance is never absolutely perfect, but in a properly operated antenna system with this type of line, we get close to the ideal.) An insulating strip or set of "crossbars" keeps the conductor spacing constant.

Coaxial Line

The cables used in community TV networks, Internet connections, and most amateur radio stations constitute examples of coaxial line. It's often called *unbalanced line* because its two conductors carry different RF currents. One conductor takes the form of a single wire running inside, and along the central axis of, a cylindrical fine-wire braid or length of metal tubing, which constitutes the other conductor. An insulating layer of solid or foamed plastic, usually *polyethylene*, keeps the conductor spacing constant. In a properly operating system using coaxial cable, the outer conductor or *shield* prevents internal EM fields from escaping, and also keeps unwanted external EM fields from getting in.

Waveguide

Waveguides find widespread use at UHF and microwave frequencies (above about 300 MHz). A typical waveguide comprises a hollow metal pipe, usually having a rectangular or circular cross section. In order to efficiently propagate an EM field, a *rectangular waveguide* must have cross-sectional dimensions measuring at least 0.5 wavelength, and preferably more than 0.7 wavelength. A *circular waveguide* should be at least 0.6 wavelength in diameter, and preferably 0.7 wavelength or more. If we want a waveguide to function well, we must prevent its interior from accumulating obstructions of any kind (such as condensation, dust, cobwebs, or insects).

Characteristic Impedance

All transmission lines exhibit an interesting property that's often, somewhat imprecisely, expressed as a form of "impedance." Engineers call it the *characteristic impedance* or *surge impedance* and symbolize it as Z_o.

We can always express Z_o as a positive real number of ohms. It's a scalar quantity, not a vector quantity. Within reasonable limits, the characteristic impedance of a coaxial or parallel-wire line remains independent of the frequency of the signals that it carries. In a waveguide, the characteristic impedance varies with the frequency.

Characteristic impedance has nothing to do with the length of a transmission line, but it does depend on the line's cross-sectional dimensions and material composition. In general, large-diameter conductors spaced close together produce low values of Z_o, while small-diameter conductors placed far apart produce high values of Z_o.

Standing Waves

In antenna systems, zones of high and low RF current or voltage often occur. These zones appear as *standing waves* that represent current or voltage maxima and minima. We call the waves "standing" because they remain in fixed positions unless some characteristic of the antenna changes. The subject of standing waves has given rise to a proliferation of articles and books, as well as some myths and misconceptions among lay people and even among some engineers.

Figure 12-15 shows three examples of standing waves on the radiating element of an RF transmitting antenna. The standing waves result from RF currents (solid lines) and voltages (dashed lines) reinforcing at fixed points called *loops*, and canceling at other fixed points called *nodes*. We see three cases here, as follows:

1. Figure 12-15A shows the pattern of standing waves on a straight antenna measuring 1/2 wavelength long and fed at the center with parallel-wire line.

2. Figure 12-15B shows the pattern of standing waves on a radiating element measuring a full wavelength long and fed at the center with parallel-wire line.

3. Figure 12-15C shows the pattern of standing waves on a radiating element measuring a full wavelength long and fed at one end with parallel-wire line.

Standing waves on antenna radiators give us no reason for concern, but standing waves on transmission lines can sometimes cause trouble. On a *lossless* (ideal) RF transmission line terminated in a load that exhibits a purely resistive impedance equal to the characteristic impedance of the line, no standing waves occur. The RF voltage E and the RF current I remain uniform all along the line, and they always exist in the ratio

$$Z_o = E/I$$

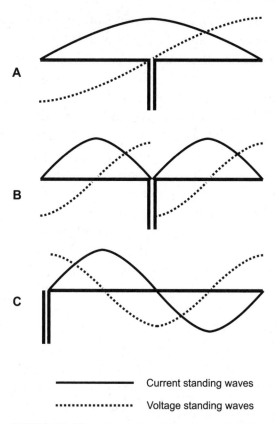

Current standing waves

Voltage standing waves

FIGURE 12-15 · Standing waves on a center-fed half-wave antenna (A), on a center-fed full-wave antenna (B), and on an end-fed full-wave antenna (C).

If the load does not exhibit a purely resistive impedance, or if it has a pure resistance that differs from the characteristic impedance of the line, an uneven distribution of current and voltage exists along the line, producing standing waves. As we increase the extent of this so-called *impedance mismatch*, the non-uniformity of current and voltage increases, and the standing waves along the line grow more pronounced.

Standing-Wave Ratio

As we move along a transmission line from one end to the other, we can measure the current and voltage at various points. We can then calculate the ratio of the maximum voltage to the minimum voltage to obtain a quantity called the *voltage standing-wave ratio* (VSWR). We can also calculate the ratio of the maximum current to the minimum current, obtaining another quantity called the *current standing-wave ratio* (ISWR). Theoretically, VSWR = ISWR regardless of the line's characteristic impedance, and regardless of the severity of the mismatch between the line and the load. Therefore, engineers commonly call either quantity the *standing-wave ratio* (SWR).

For any RF transmission line, whether lossless (the ideal case) or otherwise (the real-world case), the SWR equals 1:1, *if and only if,* the ratio of the voltage to the current is the same at every point along the line. When the SWR is 1:1, a transmission line operates at its best possible efficiency. This optimum state of affairs can exist only when we terminate a transmission line in a purely resistive load having the same ohmic value as the characteristic impedance of the line, so that the ratio of the RF voltage to the RF current equals Z_o at every point along the line from the radio to the antenna.

In practical situations, we rarely encounter a 1:1 SWR on a transmission line. In fact, we should generally expect to observe SWR values as high as 1.5:1 or even 2:1. While such mismatch levels might seem severe at first thought, they rarely cause mischief. But once we get past 2:1, line performance starts to degrade. If the SWR reaches 3:1 or more, real trouble can occur. An extremely large SWR can cause significant loss of signal power, in addition to inherent loss in the line that occurs under perfectly matched conditions. This additional loss is called *SWR loss, impedance-mismatch loss,* or *feed-line mismatch loss.* We can express, calculate, and measure SWR loss in decibels (dB), just as we can do with gain or loss in any other form.

Figure 12-16 on page 398 graphically approximates the feed-line mismatch loss in RF transmission lines for various values of matched-line loss and SWR. Engineers derive graphs of this sort by carrying out thousands of laboratory experiments with physical transmission lines, RF signal generators, and sophisticated RF power meters.

If a transmission line operates with a high SWR and the transmitter output power approaches the line's maximum rating, the peak currents and voltages can damage the line. If the current gets too high, the conductors will overheat, possibly melting the dielectric and causing a short circuit at one or more points. If the voltage gets too high, an *arc*, sometimes called *flashover*, can occur across the dielectric, permanently damaging it. For this reason, RF engineers use transmission lines capable of handling considerably more power than they endure under normal conditions. This practice allows a "safety buffer" that reduces the risk of transmission-line damage in the event something goes wrong with the antenna and drives up the SWR.

Still Struggling

A change of plus-or-minus one decibel (±1 dB) represents the smallest *sudden* increase or decrease that a listener can detect in the strength of a received signal, *if that listener anticipates the change.* If the listener does not expect the change, then the smallest *sudden* change that she can detect represents approximately ±3 dB, equivalent to a 50% increase or decrease in the signal power.

 PROBLEM 12-10

Imagine that a transmission line has 3 dB of loss when perfectly matched (SWR = 1:1). Then the situation changes, so that the SWR becomes 7:1. How much additional loss does this mismatch cause? What's the overall loss in the line when SWR equals 7:1?

 SOLUTION

Refer to Fig. 12-16. Locate the point for 3 dB on the horizontal axis. Proceed upward until you reach the curve for the 7:1 SWR. Then proceed horizontally to the left until you encounter the vertical axis. Read the SWR loss from the vertical scale. In this case it's approximately 2 dB. Therefore, the overall loss in the line equals 3 dB (the loss under perfectly matched conditions) plus 2 dB (the SWR loss), or 5 dB.

Safety Issues

Antenna systems present risks to personnel who install and maintain them. Antennas should never be placed in such a way that they can fall or blow down on power lines. Also, it should not be possible for power lines to fall or blow down on an antenna. Antennas and supporting structures should be prevented from posing a physical danger to people near them. For example, children should never have access to a radio tower because they might try to climb it.

Leave all climbing of antenna supports, such as towers and utility poles, to professionals who have received training for that sort of work, and who have sufficient insurance to cover themselves in case of an accident. You should also check your own liability insurance policy!

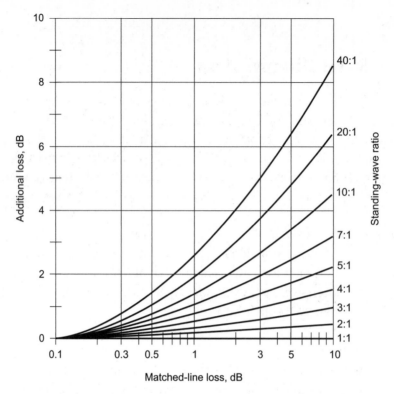

FIGURE 12-16 · Standing-wave losses produced by impedance mismatches on an RF transmission line.

Never use wireless equipment with outdoor antennas during thundershowers, or when lightning occurs anywhere in the vicinity. Do not undertake antenna construction and maintenance projects when you can see lightning, even if the storm seems far away. Disconnect antennas from electronic equipment, and connect the antennas to a substantial earth ground, except when you actually operate the equipment.

Indoor transmitting antennas expose operating personnel to EM field energy. Engineers and medical professionals have not conclusively quantified the extent of the hazard, if any, produced by such exposure. However, sufficient concern prevails to warrant checking the latest publications on the topic. For detailed information concerning antenna safety, consult a professional antenna engineer and/or a comprehensive text on antenna design and construction.

TIP *If you're not completely confident in your ability to install or maintain an antenna system, leave the job to professionals—and check out their credentials and reputation before you hire them!*

QUIZ

Refer to the text in this chapter if necessary. A good score is eight correct. The answers are in the back of the book.

1. Suppose that we want to construct a unidirectional end-fire array. Which of the following construction procedures should we follow?
   A. Place two parallel half-wave dipoles 1/4 wavelength apart and feed them 180° out of phase.
   B. Place two parallel half-wave dipoles 1/2 wavelength apart and feed them 90° out of phase.
   C. Place two parallel half-wave dipoles 1/4 wavelength apart and feed them 90° out of phase.
   D. Place two parallel half-wave dipoles 1/2 wavelength apart and feed them in phase.

2. Consider an antenna that has a radiation resistance of 73.5 ohms and a loss resistance of 26.5 ohms. What's its efficiency?
   A. 26.5%
   B. 73.5%
   C. 36.1%
   D. 47.0%

3. Suppose that we increase the radiation resistance in the scenario of Question 3 from 73.5 ohms to 735 ohms, while the loss resistance remains at 26.5 ohms. What happens to the antenna efficiency?
   A. It doesn't change.
   B. It decreases to 3.5%.
   C. It increases to 90.0%.
   D. It increases to 96.5%.

4. In order for a transmission line to exhibit an SWR of 1:1, we must connect it to a load
   A. whose resistive component equals the characteristic impedance of the line.
   B. that has no net reactance.
   C. whose absolute-value impedance equals the characteristic impedance of the line.
   D. More than one of the above

5. How can we improve the directivity of a paraboloidal dish antenna fed with a simple dipole antenna at its focal point?
   A. Use it at a higher frequency, making sure that we trim the dipole for resonance at the new frequency.
   B. Feed it with a horn antenna instead of a dipole, pointing the horn back toward the center of the reflector.
   C. Increase the reflector diameter, making sure that it maintains the optimum paraboloidal shape.
   D. Any or all of the above

6. Suppose that we acquire an old two-element Yagi antenna at an amateur-radio convention. We measure the end-to-end length of the element that connects to the feed line as 4.81 m, and the end-to-end length of the other element (which does not connect to the feed line) as 4.47 m. We can conclude that the non-feed-line element constitutes a

   A. reflector.
   B. director.
   C. phased radiator.
   D. harmonic element.

7. If we clean up the Yagi antenna described in Question 6 and decide to use it for wireless communications, on which of the following frequencies should we expect it to perform the best?

   A. 28.1 MHz
   B. 14.1 MHz
   C. 56.3 MHz
   D. 7.05 MHz

8. What's the wavelength of a 24.0-MHz wireless signal in free space?

   A. 12.5 m
   B. 5.25 m
   C. 41.0 m
   D. 9.75 m

9. What's the optimum end-to-end length for a half-wave open dipole antenna constructed from wire and intended for use at 3.535 MHz?

   A. 40.5 ft
   B. 66.1 ft
   C. 132 ft
   D. 20.3 ft

10. A full-wavelength loop antenna exhibits its maximum power gain

   A. in the plane of the loop.
   B. perpendicular to the plane of the loop.
   C. at 45° angles to the plane of the loop.
   D. in all directions; it's an isotropic radiator.

Test: Part III

Do not refer to the text when taking this test. You may draw diagrams or use a calculator if necessary. A good score is at least 38 correct. Answers are in the back of the book. It's best to have a friend check your score the first time, so you won't memorize the answers if you want to take the test again.

1. **How can we use an AM receiver to listen to an FM communications signal?**
   A. We can switch off the envelope detector.
   B. We can switch off the local oscillator.
   C. We can employ slope detection.
   D. We can tune the receiver to the signal's second harmonic.
   E. We can't.

2. **In theory, an EM field can have a frequency**
   A. anywhere between 10 kHz and 3000 GHz.
   B. anywhere between 20 Hz and 20 kHz.
   C. anywhere between 20 Hz and the red end of the visible-light spectrum.
   D. of any value more than 20 Hz.
   E. of any value more than zero.

3. **In a preamplifier, such as the type we might connect between an antenna and a superheterodyne communications receiver, nonlinearity**
   A. allows for efficient signal mixing.
   B. can cause unwanted IMD.
   C. produces excessive gain.
   D. improves the noise figure.
   E. All of the above

4. **The index of refraction in the earth's atmosphere generally decreases with increasing altitude. This phenomenon is largely responsible for radio-wave**
   A. weather-front scattering.
   B. tropospheric bending.
   C. line-of-sight propagation.
   D. ionospheric refraction.
   E. surface-wave propagation.

5. **In the Morse code, one bit represents the duration of one**
   A. character.
   B. dot.
   C. dash.
   D. word.
   E. symbol.

6. **Figure Test III-1 shows a frequency-versus-time graphical rendition of**
   A. an on/off-keyed signal modulated with a sine wave.
   B. a PDM signal modulated with a sine wave.
   C. a wideband AM signal modulated with a sine wave.
   D. a narrowband FM signal modulated with a sine wave.
   E. an HDTV signal modulated with a sine wave.

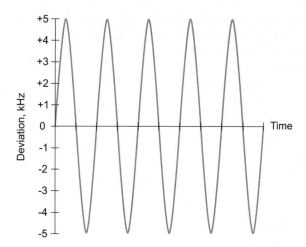

FIGURE TEST III-1 • Illustration for Part III Test Question 6.

7. Suppose that we find the instruction manual for an old "shortwave" radio receiver. The specifications page states that the receiver requires 0.5 microvolts at the antenna terminals to produce a 10-dB signal-to-noise ratio at the audio output. This information quantifies the receiver's

   A. selectivity.
   B. noise figure.
   C. gain.
   D. sensitivity.
   E. dynamic range.

8. Which type of FM demodulation circuit produces an output voltage that depends on the instantaneous signal frequency?

   A. Envelope detector
   B. Mixer
   C. Preselector
   D. Product detector
   E. Ratio detector

9. In a superheterodyne communications receiver, a low IF offers better performance than a high IF for the purpose of obtaining

   A. a high degree of selectivity.
   B. a low noise figure.
   C. a low signal-to-noise ratio.
   D. maximum dynamic range.
   E. reception free of images and birdies.

10. **Consider an EM field whose wavelength equals 1.5 km in free space. If we reduce the wavelength to 15 m, what happens to the frequency?**
    A. It increases by a factor of 100.
    B. It increases by a factor of 10.
    C. It decreases by a factor of 10.
    D. It decreases by a factor of 100.
    E. We need more information to answer this question.

11. **Which of the following demodulation circuits would work best for the reception of SSB signals?**
    A. Slope detector
    B. Phase-locked loop
    C. Envelope detector
    D. Product detector
    E. Ratio detector

12. **Figure Test III-2 is a schematic diagram of a simple receiver for AM signals. What, if any, fundamental error appears in this diagram?**
    A. The JFET should be replaced with a PNP bipolar transistor.
    B. The coil should be replaced with a potentiometer.
    C. The variable capacitor should be replaced with a variable inductor.
    D. The power-supply polarity is wrong; it should provide −12 V at the drain.
    E. Figure Test III-2 contains no fundamental errors.

13. **What's the purpose of the component marked X in the circuit of Fig. Test III-2?**
    A. Envelope detection
    B. Intermodulation prevention
    C. Noise-figure reduction

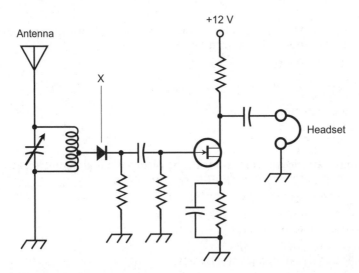

FIGURE TEST III-2 · Illustration for Part III Test Questions 12 and 13.

D. Selectivity enhancement

E. Gain improvement

14. What decimal quantity does the hexadecimal numeral D represent?

A. 10

B. 11

C. 12

D. 13

E. 14

15. What decimal quantity does the binary numeral 1111 represent?

A. 7

B. 8

C. 15

D. 16

E. 33

16. If we add 1 to the binary quantity 1111, what binary quantity do we get?

A. 1112

B. 1110

C. 10001

D. 11000

E. None of the above

17. A typical AM voice communications signal has a bandwidth of approximately

A. 300 Hz.

B. 1 kHz.

C. 3 kHz.

D. 6 kHz.

E. 20 kHz.

18. In an AM signal modulated 100%, how much of the signal power actually conveys the data that we want to transmit?

A. 1/3 of it

B. Half of it

C. 2/3 of it

D. 3/4 of it

E. All of it

19. When we use multiple communications lines or a wideband channel, sending and receiving independent sequences of bits along each line or subchannel in order to maximize the data transmission speed, we employ a technology called

A. parallel data transmission.

B. synchronous data transmission.

C. analog data transmission.

D. serial data transmission.

E. polymorphous data transmission.

20. Suppose that we place a quarter-wavelength vertical antenna element over a well-designed radial system laid on the surface of the ground. We feed the element with coaxial cable, connecting the cable's center conductor to the base of the antenna element and the cable's shield to the center of the radial system. Then we gradually reduce the frequency at which we operate the antenna. What happens to the radiation resistance at the antenna feed point as we carry out this process?

A. It does not change.
B. It decreases.
C. It increases.
D. It alternately increases and decreases.
E. We need more information to answer this question.

21. Which of the following signal codes might we reasonably expect to encounter in a frequency-shift-keyed (FSK) wireless signal?

A. PM
B. SSB
C. DSB
D. AM
E. ASCII

22. What octal numeral should we write to represent the decimal quantity 17?

A. 17
B. 19
C. 20
D. 21
E. 24

23. One megabit per second equals

A. 1,000,000 bits per second.
B. 1,048,576 bits per second.
C. 1,024 kilobits per second.
D. 10 megabytes per second.
E. 1,000 kilobytes per second.

24. In A/D conversion, we call the number of different digital output states the

A. binary code.
B. byte count.
C. bit rate.
D. baud rate.
E. sampling resolution.

25. Figure Test III-3 is a diagram of a system of computers commonly known as a

A. client-server network.
B. peer-to-peer network.
C. multifaceted network.
D. zoned network.
E. diversified network.

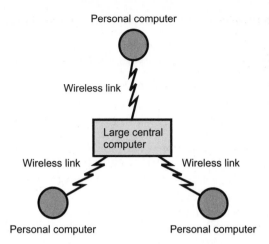

FIGURE TEST III-3 · Illustration for Part III Test Questions 25 and 26.

26. The large, central computer in the system of Fig. Test III-3 is called a
    A. diversifier.
    B. zoner.
    C. peer.
    D. server.
    E. firewall.

27. If we want a modulated RF carrier to provide effective, reliable data transfer, the highest frequency component in the modulating information should not exceed
    A. 0.1% of the carrier frequency.
    B. 1% of the carrier frequency.
    C. 10% of the carrier frequency.
    D. 50% of the carrier frequency.
    E. the carrier frequency.

28. What phenomenon can cause a nearby lightning strike to induce a destructive current surge in a wireless-communications antenna?
    A. Capacitive reactance
    B. Inductive coupling
    C. Radiation resistance
    D. Characteristic impedance
    E. Electromagnetic pulse

29. A folded-dipole antenna works well at its fundamental design frequency and
    A. all whole-numbered multiples thereof.
    B. all even-numbered multiples thereof.
    C. all odd-numbered multiples thereof.
    D. all whole-number fractions thereof.
    E. no other frequencies whatsoever.

30. **In a superheterodyne communications receiver, most of the amplification takes place in the**

    A. product-detector stage.
    B. audio-frequency (AF) stages.
    C. intermediate-frequency (IF) stages.
    D. filter stage.
    E. mixer stage.

31. **In order to reduce fading in "shortwave" radio reception that occurs when signals from distant transmitters return to the earth from the ionosphere, we can**

    A. do nothing, because it's theoretically impossible to reduce fading.
    B. use a dual-conversion receiver.
    C. employ diversity reception.
    D. install a preamplifier between the detector and the audio stages.
    E. use a low intermediate frequency.

32. **We can make a quarter-wave, ground-mounted vertical antenna function on several different frequencies by**

    A. using a feed line measuring exactly a half electrical wavelength at the fundamental operating frequency.
    B. using a network of ground radials, each measuring at least a half wavelength at the lowest intended operating frequency.
    C. using a waveguide as the feed line.
    D. inserting resonant *LC* traps at specific points along the radiating element.
    E. no known means.

33. **Figure Test III-4 is a block diagram of a**

    A. wide-area reception system.
    B. double-sideband reception system.
    C. synchronous reception system.
    D. spread-spectrum reception system.
    E. diversity reception system.

34. **A loopstick antenna exhibits the best RF signal response (that is, it's the most sensitive to incoming signals) in directions that lie**

    A. along the axis of the rod-shaped core.
    B. perpendicular to the axis of the rod-shaped core.
    C. at 45° angles to the axis of the rod-shaped core.
    D. at various angles to the axis of the rod-shaped core, depending on the frequency.
    E. anywhere; it doesn't matter because a loopstick constitutes an isotropic antenna.

35. **Suppose that in order to gain access to a computer system, you must place your thumb on a small pad that "reads" your thumb print and thereby identifies you. This technology is an example of**

    A. analog security.
    B. low-level security.

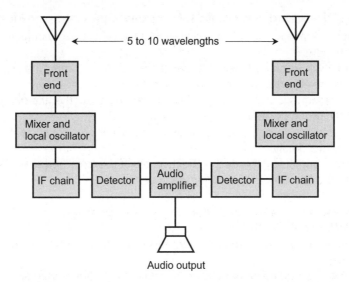

FIGURE TEST III -4 · Illustration for Part III Test Question 33.

C. biometric security.
D. optical security.
E. tactile security.

36. **Which of the following measurable parameters can represent the shape factor in a superheterodyne communications receiver?**

A. The power gain for a strong signal divided by the power gain for a weak signal
B. The intermediate-frequency (IF) filter bandwidth for 30-dB attenuation divided by its bandwidth for 3-dB attenuation
C. The noise figure in decibels minus the overall gain in decibels
D. The gain of the audio-frequency (AF) stage divided by the gain of the intermediate-frequency (IF) stage
E. The dynamic range in decibels minus the power gain in decibels

37. **At an altitude of approximately 35,800 km (22,200 mi), a satellite in a circular orbit takes one solar day to complete each revolution around the earth. If a satellite is placed in such an orbit over the equator, and if it revolves in the same direction as the earth rotates, then it**

A. appears to remain in a fixed position in the sky as viewed from any point on the surface from which we can "see" it.
B. appears to sweep across the sky gradually from east to west as viewed from any point on the surface from which we can "see" it.
C. appears to sweep across the sky gradually from west to east as viewed from any point on the surface from which we can "see" it.
D. appears to remain in a fixed position relative to the sun as viewed from any point on the surface from which we can "see" it.
E. appears to remain in a fixed position relative to the distant stars as viewed from any point on the surface from which we can "see" it.

38. **The rule "Never touch two grounds at the same time!" refers to the tendency for**

    A. ground loops to exist even when we have taken every reasonable precaution to avoid them.
    B. dangerous voltages to exist between points that common sense suggests should be electrically neutral with respect to each other.
    C. electromagnetic interference (EMI) to generate destructive and dangerous currents in ground systems.
    D. hazardous magnetic fields to arise as a result of currents flowing among various points in a circuit.
    E. human contact to degrade the effectiveness of the radio-frequency (RF) ground connection in certain types of antenna systems.

39. **A quad antenna operates according to the same principles as the Yagi antenna does, except that the quad**

    A. employs full-wavelength loop elements rather than half-wavelength straight elements.
    B. operates as a bidirectional antenna, while the Yagi functions as a unidirectional antenna.
    C. uses phased elements, while the Yagi employs parasitic elements.
    D. has a driven element and a director, while a Yagi has a driven element and a reflector.
    E. can have an unlimited number of parasitic elements, while a Yagi can never have more than two parasitic elements.

40. **Figure Test III-5 is a block diagram showing**

    A. wireless-only encryption.
    B. end-to-end encryption.
    C. strong encryption.
    D. analog encryption.
    E. dual-stage encryption.

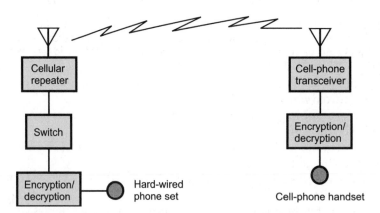

FIGURE TEST III-5 · Illustration for Part III Test Question 40.

41. Suppose that we construct a half-wave dipole antenna, and then place a straight wire near and parallel to the dipole. We don't connect the straight wire to anything, but we take care to ensure that it's a little longer than the dipole. Our goal is to obtain an antenna that exhibits power gain in certain directions. In this situation, the straight wire comprises

    A. an isotropic element.
    B. a phased element.
    C. a parasitic element.
    D. a polarizing element.
    E. an open element.

42. Figure Test III-6 is a diagram of

    A. an open dipole antenna.
    B. a zepp antenna.
    C. a Yagi antenna.
    D. a collinear antenna.
    E. an isotropic antenna.

43. The antenna shown in Fig. Test III-6 will work satisfactorily at

    A. the fundamental frequency and all even harmonics thereof.
    B. the fundamental frequency and all odd harmonics thereof.
    C. the fundamental frequency and all harmonics thereof.
    D. any frequency at which the radiating element measures some whole-number multiple of 1/2 electrical wavelength.
    E. All of the above

44. In an SSTV signal, the transmission of a vertical sync pulse

    A. regulates the contrast.
    B. transmits the color code information.
    C. tells the receiver to begin a new frame.
    D. determines the maximum brightness.
    E. defines the frame scan rate.

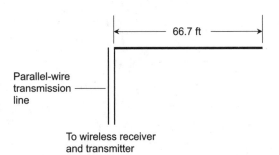

FIGURE TEST III-6 · Illustration for Part III Test Questions 42 and 43.

45. **We can use a waveguide to feed a**
    A. dipole antenna at 7 MHz.
    B. Yagi antenna at 28 MHz.
    C. helical antenna at 500 kHz.
    D. horn antenna at 15 GHz.
    E. dish antenna at 35 kHz.

46. **Suppose that we have a direct-conversion receiver and we want to receive an on/off-keyed Morse-code signal on an amateur-radio band. Which of the following actions, if any, will result in good reception?**
    A. Switch the local oscillator off.
    B. Set the local oscillator 700 Hz above the incoming signal frequency.
    C. Set the local oscillator to the second harmonic of the incoming signal frequency.
    D. Set the local oscillator precisely at the incoming signal frequency.
    E. None of the above actions will work in this situation because direct-conversion receivers cannot demodulate on/off-keyed signals.

47. **What do engineers call the extent to which a receiver can maintain a fairly constant output, while keeping its rated sensitivity in the presence of incoming signals whose amplitudes vary greatly?**
    A. Gain figure
    B. Dynamic range
    C. Attenuation gain
    D. Skirt selectivity
    E. Amplitude bandwidth

48. **Figure Test III-7 shows the H-plane directional pattern for a hypothetical antenna, as determined by field testing. Suppose that each adjacent pair of concentric circles in the coordinate system represents a difference of 5 dB in field strength or response. This pattern might reasonably represent the behavior of**

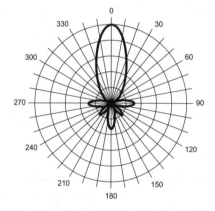

FIGURE TEST III-7. Illustration for Part III Test Question 48.

A. a half-wave open dipole antenna.

B. a half-wave folded dipole antenna.

C. a half-wave zepp antenna.

D. a Yagi antenna.

E. an isotropic antenna.

49. **Which, if any, of the following adverse effects might be observed if we operate a transmission line with a high standing-wave ratio (SWR)?**

A. Low line voltage

B. Low line current

C. Excessive power loss

D. Excessive radiation resistance

E. None of the above

50. **A voice communications signal contains audio components mostly in the range between**

A. 3 Hz and 30 Hz.

B. 30 Hz and 300 Hz.

C. 300 Hz and 3 kHz.

D. 3 kHz and 30 kHz.

E. 30 kHz and 300 kHz.

Final Exam

Do not refer to the text when taking this test. You may draw diagrams or use a calculator if necessary. A good score is at least 75 correct. You'll find the correct answers listed in the back of the book. Have a friend check your score the first time, so you won't memorize the answers if you want to take the test again.

1. Suppose that we measure the period of a particular AC wave as 1/300 of a minute. What's the frequency?

    A. 300 Hz
    B. 60.0 Hz
    C. 5.00 Hz
    D. 0.200 Hz
    E. We need more information to answer this question.

2. If we drive 300 mA through a coil of wire that contains 50 turns, how much magnetomotive force does the coil produce?

    A. 6 At
    B. 15 At
    C. 24 At
    D. 30 At
    E. We need more information to calculate it.

3. If a transformer converts 120 V RMS AC input to 30 V RMS AC output, the primary-to-secondary turns ratio equals

    A. 16:1.
    B. 4:1.
    C. 1:4.
    D. 1:16.
    E. a value that depends on the output (or load) resistance.

4. How much current flows through the resistor marked with the query symbol (?) in the circuit of Fig. Exam-1?

    A. 25 mA
    B. 33 mA
    C. 50 mA

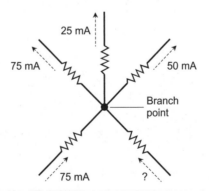

FIGURE EXAM-1 · Illustration for Final Exam Question 4.

D. 75 mA

E. We need more information to calculate it.

5. The length of a complex-impedance vector in the *RX* half-plane represents

A. admittance.

B. reactance.

C. resistance.

D. susceptance.

E. None of the above

6. Suppose that an AC wave with a large DC component has a maximum instantaneous voltage of +100 V and a minimum instantaneous voltage of +20 V. What's the peak-to-peak (pk-pk) voltage?

A. 200 V pk-pk

B. 120 V pk-pk

C. 80 V pk-pk

D. 60 V pk-pk

E. We can't define it.

7. Imagine that a high-tension power line has a loss resistance of 20.0 ohms and carries 150 A of current. How much power does the line dissipate as heat, representing power that can never reach the end users?

A. 450 kW

B. 300 kW

C. 200 kW

D. 150 kW

E. 50 kW

8. Suppose that we cut the voltage in the power transmission line described in Question 7 in half. Assume that the line resistance remains constant, but the line current doubles to 300 A. How much power does the line dissipate as heat now?

A. 200 kW

B. 600 kW

C. 900 kW

D. 1.80 MW

E. 4.00 MW

9. Consider two admittances $Y_1 = G_1 + jB_1$ and $Y_2 = G_2 + jB_2$ connected in parallel. We can express the complex admittance Y of the combination as

A. $Y = (G_1 + B_1) + j(G_2 + B_2)$.

B. $Y = (B_1 + B_2) + j(G_1 + G_2)$.

C. $Y = (G_1 + G_2) + j(B_1 + B_2)$.

D. $Y = [(G_1 + G_2)^2 + (B_1 + B_2)^2]^{1/2}$.

E. $Y = [(G_1 + G_2)^2 - (B_1 + B_2)^2]^{1/2}$.

10. How much *positive* DC voltage, at a minimum, must we superimpose on an AC wave with $E_{pk+} = +100\,V$ pk+ and $E_{pk-} = -70\,V$ pk– to obtain a fluctuating DC voltage as opposed to an alternating voltage (that is, to prevent the polarity from ever reversing)?

A. +170 V DC

B. +100 V DC

C. +70 V DC

D. +30 V DC

E. +15 V DC

11. How much *negative* DC voltage, at a minimum, must we superimpose on an AC wave with $E_{pk+} = +100\,V$ pk+ and $E_{pk-} = -70\,V$ pk– to obtain a fluctuating DC voltage as opposed to an alternating voltage (that is, to prevent the polarity from ever reversing)?

A. –170 V DC

B. –100 V DC

C. –70 V DC

D. –30 V DC

E. –15 V DC

12. Figure Exam-2 shows two pure sine waves *X* and *Y*, neither of which has any DC component. They have the same frequency, but wave *Y* occurs 180° out of phase with wave *X*. Each vertical graph division represents exactly 1 V. Based on this information, the composite that we get when we superimpose waves *X* and *Y* is a

A. sine wave of about 5.4 V pk-pk.

B. sine wave of about 2.7 V pk-pk.

C. sine wave of about 10.8 V pk-pk.

D. non-sine wave of about 7.1 V pk-pk.

E. non-sine wave of about 14.2 V pk-pk.

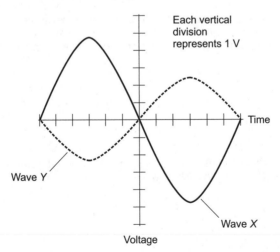

FIGURE EXAM-2 · Illustration for Final Exam Question 12.

13. **We can protect sensitive electronic equipment against the effects of sudden, dramatic spikes in the instantaneous power-supply voltage by connecting a**

    A. thyrector between each of the AC lines and electrical ground.
    B. Gunn diode between each of the AC lines and electrical ground.
    C. capacitor in series with each of the AC lines.
    D. small-value resistor in series with each of the AC lines.
    E. rectifier diode in series with each of the AC lines.

14. **If we provide a full-wave rectifier circuit with 60-Hz AC input, the output ripple frequency equals**

    A. 15 Hz.
    B. 30 Hz.
    C. 60 Hz.
    D. 120 Hz.
    E. 240 Hz.

15. **The intensity of the energy emission from an IRED depends on the forward current. As the current rises, the brightness increases up to a certain point. If the current continues to rise, no further increase in brilliance takes place. Then we say that the device has reached a state of**

    A. pinchoff.
    B. saturation.
    C. forward breakover.
    D. avalanche breakdown.
    E. negative resistance.

16. **Most people refer to electromotive force (EMF) as**

    A. current.
    B. voltage.
    C. permeability.
    D. charge.
    E. flux.

17. **Which of the following factors represents an advantage of full-wave bridge rectification over half-wave rectification, assuming that we use the same transformer in either case?**

    A. If we connect a significant load to the output of the full-wave bridge rectifier, the voltage drops less than it does with the half-wave rectifier.
    B. The full-wave bridge rectifier places less strain on the diodes than the half-wave circuit does.
    C. We can more easily filter the output of the full-wave bridge rectifier than the output of the half-wave rectifier.
    D. The full-wave bridge rectifier places less strain on the transformer than the half-wave circuit does.
    E. All of the above

18. **Which of the following electrochemical battery types has lost favor because some of its components present an environmental hazard?**

    A. Zinc-carbon
    B. Mercury
    C. Nickel-metal-hydride
    D. Alkaline
    E. Lithium

19. **In Fig. Exam-3, vector P represents**

    A. pure inductive reactance.
    B. inductive reactance combined with resistance.
    C. pure resistance.
    D. capacitive reactance combined with resistance.
    E. pure capacitive reactance.

20. **In Fig. Exam-3, vector Q represents**

    A. pure inductive reactance.
    B. inductive reactance combined with resistance.
    C. pure resistance.
    D. capacitive reactance combined with resistance.
    E. pure capacitive reactance.

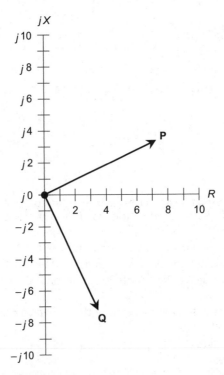

FIGURE EXAM-3 · Illustration for Final Exam
Questions 19 and 20.

21. **In an AC circuit, resistance and reactance combine mathematically to give us**

    A. susceptance.
    B. impedance.
    C. conductance.
    D. admittance.
    E. permeability.

22. **In an AC circuit, conductance and susceptance combine mathematically to give us**

    A. reactance.
    B. impedance.
    C. resistance.
    D. admittance.
    E. permittivity.

23. **A bipolar transistor can exhibit considerable dynamic current gain**

    A. below its alpha cutoff frequency.
    B. when the input signal is too weak to produce collector current during any part of the input signal cycle.
    C. in a state of saturation.
    D. above its beta cutoff frequency.
    E. All of the above

24. **In Fig. Exam-4, vector R represents**

    A. pure inductive susceptance.
    B. inductive susceptance combined with conductance.
    C. pure conductance.
    D. capacitive susceptance combined with conductance.
    E. pure capacitive susceptance.

25. **In Fig. Exam-4, vector S represents**

    A. pure inductive susceptance.
    B. inductive susceptance combined with conductance.
    C. pure conductance.
    D. capacitive susceptance combined with conductance.
    E. pure capacitive susceptance.

26. **Imagine that we feed a loudspeaker with an AF sine-wave signal at a frequency of 800 Hz. Then we suddenly increase the RMS signal voltage, keeping it at the same frequency and making sure that it retains its sine-wave nature. If the speaker impedance remains constant, what's the smallest extent of signal increase that a listener can detect if she *does not* anticipate the change?**

    A. +1 dB
    B. +2 dB
    C. +3 dB
    D. +6 dB
    E. It depends on the initial RMS signal voltage.

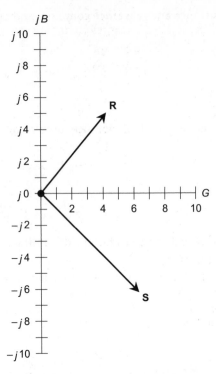

FIGURE EXAM-4 · Illustration for Final Exam
Questions 24 and 25.

27. If we connect a battery and a resistor in series with a semiconductor diode so that the positive terminal goes to the cathode and the negative terminal goes to the anode, we *always* have a condition of

   A. avalanche breakdown.
   B. forward breakover.
   C. reverse bias.
   D. rectification.
   E. saturation.

28. Figure Exam-5 is a schematic diagram of a

   A. full-wave center-tap rectifier.
   B. full-wave bridge rectifier.
   C. half-wave center-tap rectifier.
   D. half-wave bridge rectifier.
   E. voltage-doubler rectifier.

29. In an amplifier circuit that uses a JFET, we should generally try to avoid

   A. letting the drain current fluctuate when we apply an input signal.
   B. reverse-biasing the source-gate junction.

FIGURE EXAM-5 · Illustration for Final Exam Question 28.

   C. allowing current to flow through the source-gate junction under no-signal conditions.

   D. operating the device in the straight-line portion of the drain-current versus gate-voltage curve.

   E. Any of the above circumstances

30. **In a wireless communications receiver, we can connect a pair of diodes in reverse parallel with variable reverse bias across the audio output terminals to obtain**

   A. digital modulation.

   B. envelope detection.

   C. signal mixing.

   D. noise limiting.

   E. notch filtering.

31. **If we cut in half the RMS voltage of an RF signal across a pure resistance of 100 ohms, we produce a voltage-decibel change of**

   A. −8 dB.

   B. −6 dB.

   C. −4 dB.

   D. −2 dB.

   E. −1 dB.

32. **Figure Exam-6 is a schematic diagram of**

   A. a product detector.

   B. a double balanced mixer.

   C. an envelope detector.

   D. a balanced modulator.

   E. a ratio detector.

33. **If we encounter an old engineering paper that refers to "insulated-gate FETs," we can assume that the authors were writing about**

   A. JFETs.

   B. bipolar transistors.

   C. Zener diodes.

   D. PIN diodes.

   E. None of the above

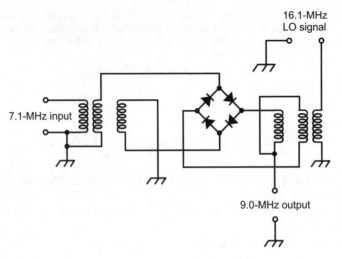

FIGURE EXAM-6 · Illustration for Final Exam Question 32.

34. **We get better results if we take the output of an RF oscillator (such as the Hartley or Colpitts type) from the emitter or source circuit rather than from the collector or drain circuit. This practice helps to ensure**

 A. optimum stability.
 B. ease of frequency adjustment.
 C. excellent linearity.
 D. maximum sensitivity.
 E. minimum feedback.

35. **The waveform produced by an audio oscillator, and therefore, the timbre of the sound that comes from a speaker or headset connected to it, depends *entirely* on the**

 A. values of the resistors and capacitors in the circuit.
 B. fundamental frequency of the crystal used in the circuit.
 C. energy at harmonic frequencies relative to the energy at the fundamental frequency.
 D. phase relationship between the input and output signals in the active device (such as a bipolar transistor or JFET).
 E. amount and type of reactance in the load.

36. **The shortest possible element in a binary frequency-shift-keyed (FSK) signal is a**

 A. dash.
 B. mark.
 C. word.
 D. character.
 E. bit.

37. Assuming proper operation, in which of the following amplifier types does the input signal *never*, even for the briefest instant, drive a bipolar transistor outside of the straight-line portion of its collector-current (I_C) versus base-current (I_B) curve?

    A. Class A
    B. Class AB$_1$
    C. Class AB$_2$
    D. Class B
    E. Class C

38. We can get decent results in a single-sideband (SSB) wireless transmitter if we operate the final power amplifier (PA) in any of the following modes except one. Which one?

    A. Class A
    B. Class AB$_1$
    C. Class AB$_2$
    D. Class B
    E. Class C

39. If we want to build an effective SSB wireless transmitter, we must

    A. maximize the bandwidth.
    B. suppress the carrier.
    C. differentiate between mark and space.
    D. amplify both sidebands equally.
    E. ensure a linear relation between frequency and phase.

40. Figure Exam-7 shows an NPN bipolar transistor connected to a 9.0-V battery and a 1.5-V cell. Under these conditions, we should expect the device to operate in a condition of

    A. cutoff.
    B. forward breakover.
    C. saturation.
    D. class-A bias.
    E. negative resistance.

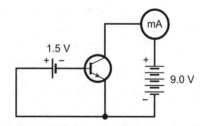

FIGURE EXAM-7 · Illustration for Final Exam Questions 40 and 41.

41. In the circuit of Fig. Exam-7, we should expect that the meter will show a current of approximately

    A. 0 mA.
    B. 150 mA.
    C. 300 mA.
    D. 900 mA.
    E. 1350 mA.

42. Imagine that we have nine silicon PV cells, each of which can produce 0.6 V at 90 mA in direct sunlight. If we connect all nine of these cells in series and then expose them to direct sunlight, and if none of the cells has enough internal resistance to affect the voltage or current from any combination of devices, we should expect a maximum deliverable current of

    A. 10 mA.
    B. 30 mA.
    C. 90 mA.
    D. 270 mA.
    E. 810 mA.

43. In a dry cell, the chemical energy, which ultimately becomes DC electricity, is contained in the

    A. carbon granules.
    B. cathode.
    C. anode.
    D. electrolyte.
    E. semiconductor material.

44. An interlaced HDTV image offers improved resolution but can also

    A. cause problems with the reproduction of fast-moving scenes.
    B. increase the bandwidth of the transmitted signal.
    C. increase the susceptibility of the system to noise and interference.
    D. inflate the overall system cost.
    E. All of the above

45. Which of the following ionospheric layers offers no advantage for high-frequency long-distance radio communication, and in fact, can have a detrimental effect?

    A. The F_2 layer, which occurs at the highest altitude
    B. The F_1 layer, which occurs somewhat below the F_2 layer but above the E layer
    C. The E layer, which occurs below the F_1 layer and above the D layer
    D. The D layer, which occurs at the lowest altitude
    E. The A layer, which can occur at any altitude

46. Which of the following circuit arrangements would work best as the basis for an RF oscillator?

    A. Common-gate
    B. Class-C

C. Common-base
D. Common-emitter
E. Class-B

47. **In theory, we can demodulate an FM signal using**
    A. an AM receiver and slope detection.
    B. a ratio detector.
    C. a discriminator and limiter.
    D. a phase-locked loop (PLL).
    E. Any of the above

48. **In a P-channel JFET, majority carriers (holes) in the channel flow between the**
    A. emitter and the collector.
    B. emitter and the source.
    C. gate and the collector.
    D. gate and the drain.
    E. source and the drain.

49. **Figure Exam-8 illustrates an example of**
    A. an amplifier using an op amp, in which the output occurs in phase coincidence with the input.
    B. an oscillator using an op amp, capable of generating signals at a certain fundamental frequency and all even harmonics thereof.
    C. an amplifier using an op amp, in which the output occurs in phase opposition relative to the input.
    D. an oscillator using an op amp, capable of generating signals at a certain fundamental frequency and all odd harmonics thereof.
    E. an audio notch filter using an op amp, in which the notch frequency can be continuously adjusted.

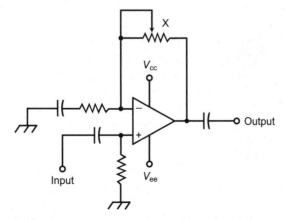

FIGURE EXAM-8 · Illustration for Final Exam Questions 49 and 50.

50. **In the circuit of Fig. Exam-8, what function does the potentiometer marked X serve?**
    A. Adjust the gain of the amplifier
    B. Adjust the frequency of oscillation
    C. Adjust the frequency of the audio notch
    D. Vary the phase of the output signal
    E. Vary the waveform of the output signal

51. **The sensitivity of a weak-signal RF amplifier depends on the**
    A. input impedance and the power output.
    B. gain and the noise figure.
    C. efficiency and the selectivity.
    D. bandwidth and the modulation index.
    E. type of detection method used.

52. **In order to produce AF energy from an incoming SSB signal and the output of a local oscillator (LO), a passive product detector takes advantage of**
    A. the gain in a bipolar or field-effect transistor.
    B. intermediate-frequency (IF) selectivity.
    C. the fluctuating feedback voltage in a PLL.
    D. the skirts of a filter response curve.
    E. the nonlinearity in semiconductor diodes.

53. **If the reactance is many times the resistance in a series *RL* circuit, then**
    A. the current and the voltage are almost in phase.
    B. the current leads the voltage by almost 1/4 cycle.
    C. the current lags the voltage by almost 1/4 cycle.
    D. the current leads the voltage by almost 1/2 cycle.
    E. the current lags the voltage by almost 1/2 cycle.

54. **If the reactance is a tiny fraction of the resistance in a series *RL* circuit, then**
    A. the current and the voltage are almost in phase.
    B. the current leads the voltage by almost 1/4 cycle.
    C. the current lags the voltage by almost 1/4 cycle.
    D. the current leads the voltage by almost 1/2 cycle.
    E. the current lags the voltage by almost 1/2 cycle.

55. **In a wireless communications receiver, the noise figure specification attains increasing importance as we**
    A. decrease the frequency.
    B. increase the frequency.
    C. increase the bandwidth.
    D. decrease the bandwidth.
    E. increase the modulation percentage.

56. Suppose that we connect a 470-ohm resistor in *series* with a 220-ohm resistor. What's the net resistance of the combination?

   A. 150 ohms
   B. 250 ohms
   C. 322 ohms
   D. 345 ohms
   E. 690 ohms

57. Suppose that we connect a 470-ohm resistor in *parallel* with a 220-ohm resistor. What's the net resistance of the combination?

   A. 150 ohms
   B. 250 ohms
   C. 322 ohms
   D. 345 ohms
   E. 690 ohms

58. If we connect a 12-V battery across the parallel combination of resistors described in Question 57, how much current does the battery drive through the combination?

   A. 35 mA
   B. 48 mA
   C. 80 mA
   D. 21 A
   E. 27 A

59. A typical variable-frequency oscillator (VFO) is designed to operate

   A. using a quartz crystal.
   B. into a high-impedance, purely resistive load.
   C. using an external frequency standard.
   D. with a phase-locked loop (PLL).
   E. as a complete RF transmitter all by itself.

60. The standard unit of magnetic flux density is the

   A. ampere-turn.
   B. gilbert.
   C. volt per ampere.
   D. tesla.
   E. coulomb.

61. Which of the following oscillator types would offer the best frequency stability if designed to operate at 10 MHz?

   A. PLL synthesizer
   B. Pierce
   C. Reinartz
   D. Hartley
   E. Clapp

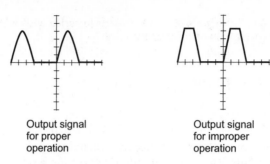

Output signal
for proper
operation

Output signal
for improper
operation

FIGURE EXAM-9 · Illustration for Final Exam Questions
63 and 64.

62. **The intrinsic layer in a PIN diode**
    A. minimizes the junction capacitance.
    B. minimizes the avalanche voltage.
    C. maximizes the forward-breakover voltage.
    D. enhances the nonlinearity.
    E. All of the above

63. **Figure Exam-9 shows output signals from an RF power amplifier that receives a sine-wave signal input. The graph at left shows the amplitude-versus-time display (as we might see it on an oscilloscope) of the output signal for proper operation. Exactly half of the waveform appears; the other half is cut off. The graph at right shows the output signal from the same amplifier when it's improperly operated. From these graphs, we can surmise that this RF power amplifier is meant to work in**
    A. class A.
    B. class AB_1.
    C. class AB_2.
    D. class B.
    E. class C.

64. **In the situation shown by the right-hand drawing in Fig. Exam-9, it appears that the amplifier is operating in a state of**
    A. push-pull.
    B. overdrive.
    C. linearity.
    D. phase opposition.
    E. impedance mismatch.

65. **Figure Exam-10 is a schematic diagram of a Hartley oscillator using a PNP bipolar transistor. What, if anything, is wrong?**
    A. The variable capacitor should go between the coil tap and ground, not across the entire coil.
    B. The variable capacitor should go between the bottom end of the coil and ground, not in parallel with the coil.

    C. The capacitor between the +13.5 V power supply connection and ground should be replaced by an RF choke.

    D. The power-supply polarity is wrong; it should be a negative DC voltage, not a positive DC voltage.

    E. Nothing is wrong with this circuit as shown.

66. **Which of the following measures offers the maximum possible amount of protection for sensitive electronic equipment during a thunderstorm?**

    A. Use a power supply that has a fuse or circuit breaker.

    B. Ensure that the power supply has adequate voltage regulation.

    C. Employ a transient suppressor in the AC utility line.

    D. Use the equipment with a 234-V system rather than a 117-V system.

    E. Unplug all equipment from the wall outlets until the storm passes.

67. **If we design a class-A JFET amplifier circuit properly, small changes in the instantaneous gate voltage, caused by an AC input signal, produce large fluctuations in the**

    A. applied DC drain voltage.

    B. instantaneous drain current.

    C. output impedance.

    D. forward breakover threshold.

    E. harmonic content.

68. **What do we call a substance that dilates magnetic flux compared with the flux density in a vacuum?**

    A. Diamagnetic

    B. Ferromagnetic

    C. Quasimagnetic

    D. Nonmagnetic

    E. Hypomagnetic

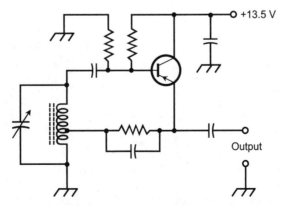

FIGURE EXAM-10 · Illustration for Final Exam Question 65.

69. In an on/off-keyed Morse-code signal, the key-up (off) condition is called
    A. dash.
    B. dot.
    C. mark.
    D. space.
    E. high.

70. Figure Exam-11 is a schematic diagram of a Reinartz oscillator using a P-channel JFET. Note the three components marked X, Y, and Z. Which of these components determines the *fundamental* oscillation frequency?
    A. X only
    B. Y only
    C. Z only
    D. Y and Z
    E. X, Y, and Z

71. Fill in the blank to make the following sentence true: "In standard AM voice communications, a signal requires, at the minimum, a range of audible frequencies from about _____ to allow reasonable intelligibility at the output of a receiver."
    A. 20 to 200 Hz
    B. 2 to 20 kHz
    C. 300 to 3000 Hz
    D. 3 to 30 MHz
    E. 0 to 5 kHz

72. If we want to receive a frequency-shift-keyed (FSK) signal using a direct-conversion receiver, we should set the local-oscillator (LO) frequency
    A. a few hundred hertz above or below both the mark and space frequencies.
    B. precisely at the mark frequency.
    C. precisely at the space frequency.

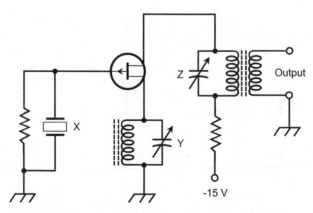

FIGURE EXAM-11 · Illustration for Final Exam Question 70.

    D. midway between the mark and space frequencies.

    E. nowhere because we can't use a direct-conversion receiver to receive an FSK signal.

73. **Frequency hopping constitutes a method of getting a transmitter and receiver to operate in**

    A. FM mode.

    B. SSB mode.

    C. DSP mode.

    D. diversity mode.

    E. spread-spectrum mode.

74. **In a set of resistors connected in parallel, the voltage across any particular resistor equals the voltage across any other resistor, and the total wattage dissipated by the entire combination equals**

    A. the wattage dissipated by any one of the individual resistors.

    B. the average of the wattages dissipated in the individual resistors.

    C. the sum of the wattages dissipated in the individual resistors.

    D. the product of the voltage across any particular resistor and the current through that resistor.

    E. the voltage across any particular resistor divided by the current through that resistor.

75. **Figure Exam-12 is a graphical rendition of**

    A. pulse-code modulation.

    B. pulse-interval modulation.

    C. pulse-width modulation.

    D. pulse-amplitude modulation.

    E. pulse-frequency modulation.

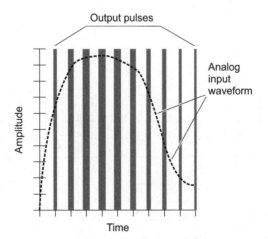

FIGURE EXAM-12 · Illustration for Final Exam Question 75.

76. **The numeral FF in hexadecimal notation represents the decimal quantity**
    A. 99.
    B. 100.
    C. 255.
    D. 999.
    E. 1001.

77. **When you use the Web to access information from a remote site, you should try to avoid**
    A. working hours at the remote location.
    B. nighttime hours at your location.
    C. nighttime hours at the remote location.
    D. 8:00 A.M. to 5:00 P.M. Coordinated Universal Time.
    E. supper time at the remote location.

78. **Figure Exam-13 is a block diagram of a direct-conversion receiver. What label should we put in the box marked X to complete this diagram?**
    A. Transient suppressor
    B. Local oscillator
    C. Product detector
    D. Frequency multiplier
    E. Programmable divider

79. **We could effectively use the system shown in Fig. Exam-13 to receive**
    A. on/off-keyed Morse-code signals.
    B. frequency-shift-keyed (FSK) signals.
    C. single-sideband (SSB) signals.
    D. All of the above
    E. None of the above

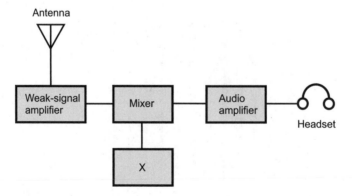

FIGURE EXAM-13 · Illustration for Final Exam Questions 78 and 79.

80. Imagine that we take two identical superheterodyne receivers and remove all of the AF stages from them both. Then we connect the detector outputs to a single AF-amplifier chain. We connect each receiver to its own antenna, and separate the antennas by several wavelengths. Finally, we set the two receivers to exactly the same frequency, so that they intercept the same signal from the "airwaves." This action represents an example of

    A. time-division multiplexing.
    B. frequency-division multiplexing.
    C. wavelength-division multiplexing.
    D. spread-spectrum reception.
    E. diversity reception.

81. In data storage, 1 TB represents

    A. 2^{40} bytes.
    B. 10^{12} bytes.
    C. 10^{6} MB.
    D. 10^{12} MB.
    E. 2^{60} MB.

82. If you want to use an amateur-radio transmitter in a motor vehicle, you should connect the radio's power cord directly to the vehicle's battery, and not through any "cigarette lighter" adapter. Why?

    A. The voltage at the "cigarette lighter" adapter is too high for the radio.
    B. If you connect the power cord through the "cigarette lighter" adapter, you run the risk of having the radio transmitter interfere with the vehicle's micro-computer.
    C. The "cigarette lighter" adapter will deliver too much current to the radio.
    D. The "cigarette lighter" adapter provides AC, not DC; if you attempt to use it, you could damage your radio.
    E. The premise is wrong. You should use the "cigarette lighter" adapter, and never connect a radio directly to the vehicle's battery.

83. In a Web site's URL, the letters "http" stand for

    A. hypertext transfer protocol.
    B. hypertext transport preference.
    C. hourly transmission to production.
    D. hop-to-transfer percentage.
    E. nothing, because we'll never see the letters "http" in a Web site's URL.

84. Figure Exam-14 is a simplified block diagram of

    A. a data buffer.
    B. a modem.
    C. a repeater.
    D. an encryption device.
    E. a serial-to-parallel interface.

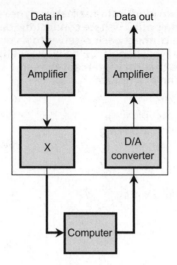

FIGURE EXAM-14 · Illustration for Final Exam Questions 84 and 85.

85. **In the system of Fig. Exam-14, box X represents**

    A. a parallel-to-serial interface.
    B. a terminal node controller.
    C. a local oscillator.
    D. an analog-to-digital converter.
    E. a mixer.

86. **When visible-light energy of varying intensity strikes the P-N junction of a reverse-biased silicon photodiode, the diode current follows along with the instantaneous fluctuations in the illumination level. This effect makes silicon photodiodes especially useful for**

    A. regulating the output of a diode oscillator.
    B. generating electrical power from the sun.
    C. receiving modulated-light signals.
    D. producing a beam of coherent light.
    E. controlling the wavelength of the diode's output.

87. **A counterpoise provides a way to get**

    A. an excellent DC ground with a single wire.
    B. an excellent RF ground without a direct connection to the earth.
    C. an antenna to function on harmonics of its fundamental frequency.
    D. a high value of radiation resistance for a short antenna.
    E. an antenna to exhibit a perfect impedance match to a transmission line over a wide range of frequencies.

88. Figure Exam-15 illustrates the basic design of a folded dipole. What straight-line distance does *L* represent at the fundamental frequency of this antenna?

    A. 0.1 electrical wavelength or less
    B. 0.25 electrical wavelength
    C. 0.5 electrical wavelength
    D. A full electrical wavelength
    E. Any whole-number multiple of a full electrical wavelength

89. When gain occurs in the direction going from the parasitic element toward the driven element in a quad antenna, we call the parasitic element a

    A. reflector.
    B. director.
    C. phased array.
    D. collinear array.
    E. waveguide.

90. We can expect a well-designed, straight zepp antenna in free space to operate at a specific fundamental frequency and

    A. only the even harmonics thereof.
    B. only the odd harmonics thereof.
    C. all the harmonics thereof.
    D. no other fundamental frequencies whatsoever.
    E. all frequencies greater than half and less than twice the fundamental frequency.

91. Which of the following data conversion methods involves the reception of bits one by one from a single line or channel, and their retransmission in batches along several lines or channels?

    A. Analog-to-digital (A/D)
    B. Digital-to-analog (D/A)
    C. Parallel-to-serial (P/S)
    D. Serial-to-parallel (S/P)
    E. Decimal-to-binary (D/B)

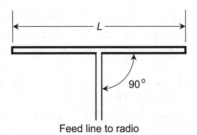

Feed line to radio

FIGURE EXAM-15 · Illustration for Final Exam Question 88.

92. Fill in the blank in the following sentence to make it true: "Radio waves travel approximately _____ as fast along a typical antenna element as they do through a vacuum."

    A. 10% to 15%
    B. 30% to 40%
    C. 50%
    D. 70% to 75%
    E. 90% to 95%

93. Suppose that we want to build a unidirectional antenna for use at 21 MHz. We want to be able to physically rotate the antenna to choose the direction of maximum radiation and response. We want the antenna to exhibit forward gain of several decibels with respect to a half-wave open dipole. Which of the following designs represents the best choice?

    A. Longwire
    B. Zepp
    C. Yagi
    D. Dish
    E. Isotropic

94. At which, if any, of the following wavelengths might we use a horn antenna?

    A. 500 m
    B. 50 m
    C. 5 m
    D. 5 cm
    E. None of the above

95. If a geostationary satellite's orbit decays so that its orbit is too low (an altitude less than approximately 35,800 km or 22,200 mi), it will appear to

    A. "wobble" around at random in the sky.
    B. drift in the sky from west to east.
    C. completely disappear from view.
    D. follow an orbit that takes it over the geographic poles.
    E. oscillate in the sky along a north-south path.

96. We define the efficiency of an RF power amplifier as the ratio of the

    A. output power with no signal input to the output power with full signal input.
    B. DC input power with no signal input to the DC input power with full signal input.
    C. resistive component of the output impedance to the resistive component of the input impedance.
    D. power output at the fundamental frequency to the total power output at the fundamental frequency and all harmonics.
    E. useful signal power output to the total DC power input.

97. What's the frequency (in conventional terms) of a sine wave whose angular frequency equals 2,261,947 rad/s? Consider $\pi = 3.14159265$.

    A. 720 kHz
    B. 7.1 MHz

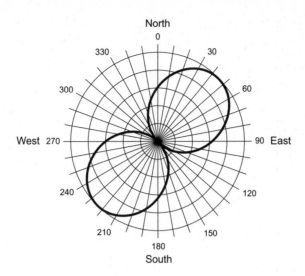

FIGURE EXAM-16 · Illustration for Final Exam Question 98.

C. 14.2 MHz
D. 36 kHz
E. 360 kHz

98. **Figure Exam-16 is a horizontal-plane (H-plane) directional graph for a certain antenna. Which type of antenna can exhibit this sort of pattern?**

    A. Folded dipole
    B. Yagi
    C. Corner reflector
    D. Helical
    E. Dish

99. **Which of the following antenna types would we most likely find at a highway call station?**

    A. Open dipole
    B. Folded dipole
    C. Zepp
    D. Ground-mounted vertical
    E. Corner reflector

100. **In which of the following amplifier types does the active device (such as a bipolar transistor or JFET) conduct current for the smallest proportion of the input-signal waveform?**

    A. Class A
    B. Class AB_1
    C. Class AB_2
    D. Class B
    E. Class C

Answers to Quizzes, Tests, and Final Exam

Chapter 1	Chapter 3	Test: Part I	22. A
1. B	1. C	1. B	23. C
2. B	2. B	2. D	24. D
3. C	3. A	3. C	25. D
4. A	4. D	4. A	26. C
5. D	5. D	5. A	27. C
6. D	6. C	6. E	28. E
7. A	7. C	7. D	29. B
8. D	8. B	8. E	30. D
9. D	9. A	9. B	31. C
10. C	10. C	10. D	32. E
		11. C	33. A
Chapter 2	Chapter 4	12. C	34. D
1. B	1. C	13. B	35. C
2. B	2. A	14. B	36. A
3. C	3. B	15. B	37. B
4. B	4. B	16. D	38. D
5. D	5. D	17. A	39. E
6. A	6. A	18. B	40. C
7. C	7. C	19. A	41. E
8. D	8. D	20. D	42. A
9. C	9. B	21. E	43. B
10. D	10. A		

44. B
45. A
46. C
47. A
48. D
49. A
50. D

Chapter 5
1. D
2. C
3. B
4. A
5. B
6. C
7. A
8. D
9. C
10. B

Chapter 6
1. A
2. B
3. D
4. B
5. C
6. B
7. C
8. C
9. D
10. A

Chapter 7
1. D
2. A
3. A
4. D
5. B
6. D

7. B
8. C
9. A
10. C

Chapter 8
1. D
2. C
3. A
4. D
5. C
6. B
7. C
8. B
9. A
10. A

Test: Part II
1. C
2. B
3. B
4. C
5. C
6. E
7. A
8. E
9. E
10. A
11. D
12. B
13. C
14. D
15. B
16. E
17. D
18. C
19. B
20. C
21. D

22. A
23. D
24. C
25. D
26. B
27. E
28. D
29. B
30. A
31. B
32. E
33. E
34. D
35. B
36. A
37. A
38. C
39. E
40. B
41. A
42. C
43. D
44. A
45. D
46. A
47. E
48. B
49. C
50. D

Chapter 9
1. C
2. C
3. D
4. B
5. B
6. A
7. B

8. D
9. A
10. D

Chapter 10
1. C
2. B
3. B
4. B
5. D
6. D
7. A
8. D
9. C
10. A

Chapter 11
1. D
2. A
3. B
4. B
5. A
6. C
7. D
8. D
9. C
10. D

Chapter 12
1. C
2. B
3. D
4. D
5. D
6. B
7. A
8. A
9. C
10. B

Test: Part III	38. B	25. B	63. D
1. C	39. A	26. C	64. B
2. E	40. B	27. C	65. D
3. B	41. C	28. A	66. E
4. B	42. B	29. C	67. B
5. B	43. E	30. D	68. A
6. D	44. C	31. B	69. D
7. D	45. D	32. B	70. A
8. E	46. B	33. E	71. C
9. A	47. B	34. A	72. A
10. A	48. D	35. C	73. E
11. D	49. C	36. E	74. C
12. E	50. C	37. A	75. C
13. A		38. E	76. C
14. D	Final Exam	39. B	77. A
15. C	1. C	40. A	78. B
16. E	2. B	41. A	79. D
17. D	3. B	42. C	80. E
18. A	4. D	43. D	81. A
19. A	5. E	44. A	82. B
20. B	6. C	45. D	83. A
21. E	7. A	46. D	84. B
22. D	8. D	47. E	85. D
23. A	9. C	48. E	86. C
24. E	10. C	49. A	87. B
25. A	11. B	50. A	88. C
26. D	12. A	51. B	89. A
27. C	13. A	52. E	90. C
28. E	14. D	53. C	91. D
29. C	15. B	54. A	92. E
30. C	16. B	55. B	93. C
31. C	17. E	56. E	94. D
32. D	18. B	57. A	95. B
33. E	19. B	58. C	96. E
34. B	20. D	59. B	97. E
35. C	21. B	60. D	98. A
36. B	22. D	61. A	99. E
37. A	23. A	62. A	100. E
	24. D		

Schematic Symbols

ammeter		antenna, loop, multiturn	
amplifier, general		battery, electrochemical	
amplifier, inverting		capacitor, feedthrough	
amplifier, operational		capacitor, fixed	
		capacitor, variable	
AND gate		capacitor, variable, split-rotor	
antenna, balanced			
antenna, general		capacitor, variable, split-stator	
antenna, loop			

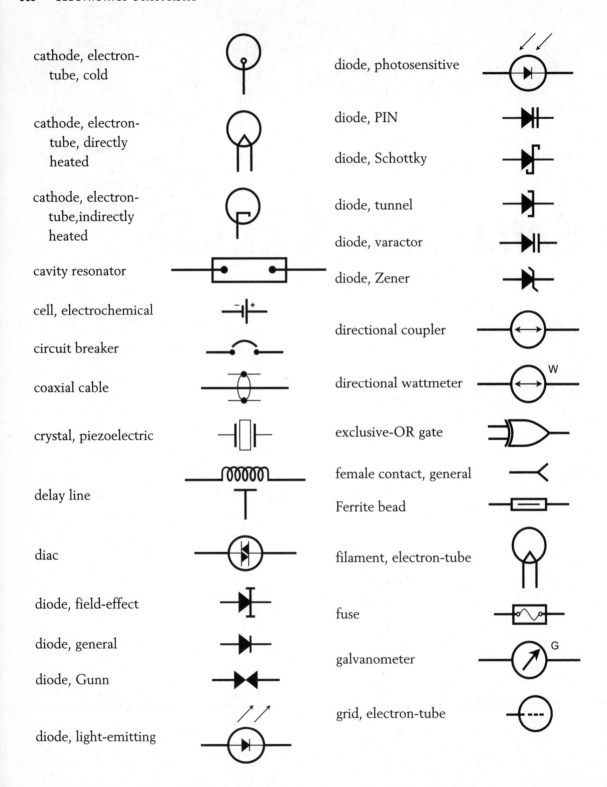

cathode, electron-tube, cold

cathode, electron-tube, directly heated

cathode, electron-tube, indirectly heated

cavity resonator

cell, electrochemical

circuit breaker

coaxial cable

crystal, piezoelectric

delay line

diac

diode, field-effect

diode, general

diode, Gunn

diode, light-emitting

diode, photosensitive

diode, PIN

diode, Schottky

diode, tunnel

diode, varactor

diode, Zener

directional coupler

directional wattmeter

exclusive-OR gate

female contact, general

Ferrite bead

filament, electron-tube

fuse

galvanometer

grid, electron-tube

ground, chassis	
ground, earth	
handset	
headset, double	
headset, single	
headset, stereo	L R
inductor, air core	
inductor, air core, bifilar	
inductor, air core, tapped	
inductor, air core, variable	
inductor, iron core	
inductor, iron core, bifilar	
inductor, iron core, tapped	

inductor, iron core, variable	
inductor, powdered-iron core	
inductor, powdered-iron core, bifilar	
inductor, powdered-iron core, tapped	
inductor, powdered-iron core, variable	
integrated circuit, general	(Part No.)
jack, coaxial or phono	
jack, phone, 2-conductor	
jack, phone, 3-conductor	
key, telegraph	
lamp, incandescent	
lamp, neon	
male contact, general	

meter, general		outlet, 234-volt	
microammeter	μA	plate, electron-tube	
microphone		plug, 2-wire, nonpolarized	
microphone, directional		plug, 2-wire, polarized	
milliammeter	mA	plug, 3-wire	
NAND gate		plug, 234-volt	
negative voltage connection	−	plug, coaxial or phono	
NOR gate		plug, phone, 2-conductor	
NOT gate		plug, phone, 3-conductor	
optoisolator		positive voltage connection	+
OR gate		potentiometer	
outlet, 2-wire, nonpolarized		probe, radio-frequency	or
outlet, 2-wire, polarized			
outlet, 3-wire			

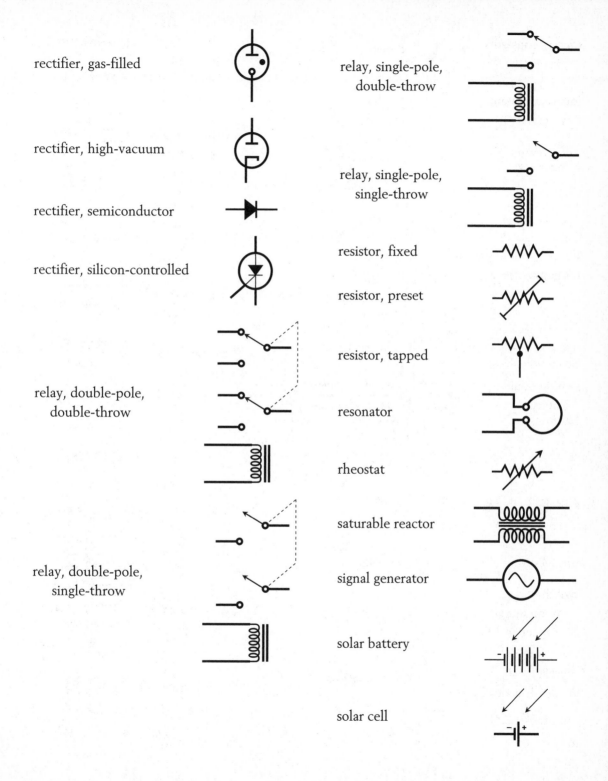

rectifier, gas-filled

rectifier, high-vacuum

rectifier, semiconductor

rectifier, silicon-controlled

relay, double-pole, double-throw

relay, double-pole, single-throw

relay, single-pole, double-throw

relay, single-pole, single-throw

resistor, fixed

resistor, preset

resistor, tapped

resonator

rheostat

saturable reactor

signal generator

solar battery

solar cell

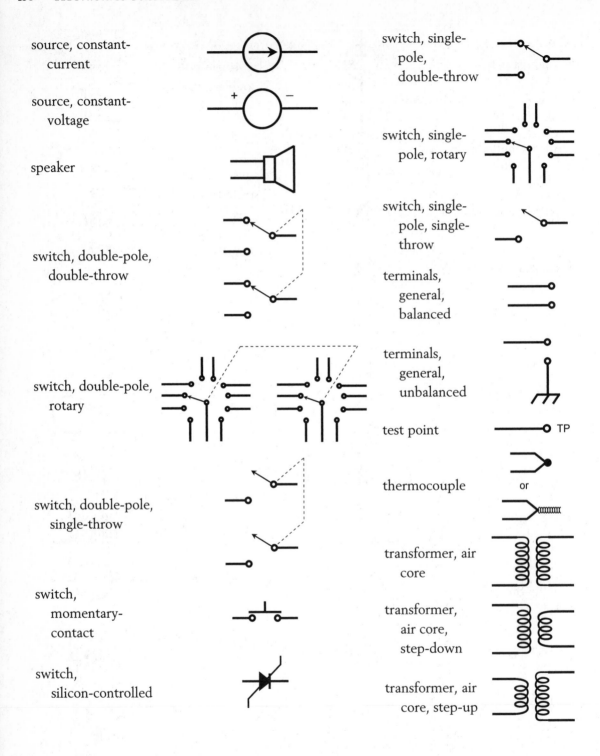

source, constant-current	
source, constant-voltage	
speaker	
switch, double-pole, double-throw	
switch, double-pole, rotary	
switch, double-pole, single-throw	
switch, momentary-contact	
switch, silicon-controlled	
switch, single-pole, double-throw	
switch, single-pole, rotary	
switch, single-pole, single-throw	
terminals, general, balanced	
terminals, general, unbalanced	
test point	TP
thermocouple	or
transformer, air core	
transformer, air core, step-down	
transformer, air core, step-up	

transformer, air core, tapped primary		transformer, powdered-iron core, tapped secondary	
transformer, air core, tapped secondary		transistor, bipolar, NPN	
transformer, iron core		transistor, bipolar, PNP	
transformer, iron core, step-down		transistor, field-effect, N-channel	
transformer, iron core, step-up		transistor, field-effect, P-channel	
transformer, iron core, tapped primary		transistor, MOS field-effect, N-channel	
transformer, iron core, tapped secondary		transistor, MOS field-effect, P-channel	
transformer, powdered-iron core		transistor, photosensitive, NPN	
transformer, powdered-iron core, step-down		transistor, photosensitive, PNP	
transformer, powdered-iron core, step-up			
transformer, powdered-iron core, tapped primary			

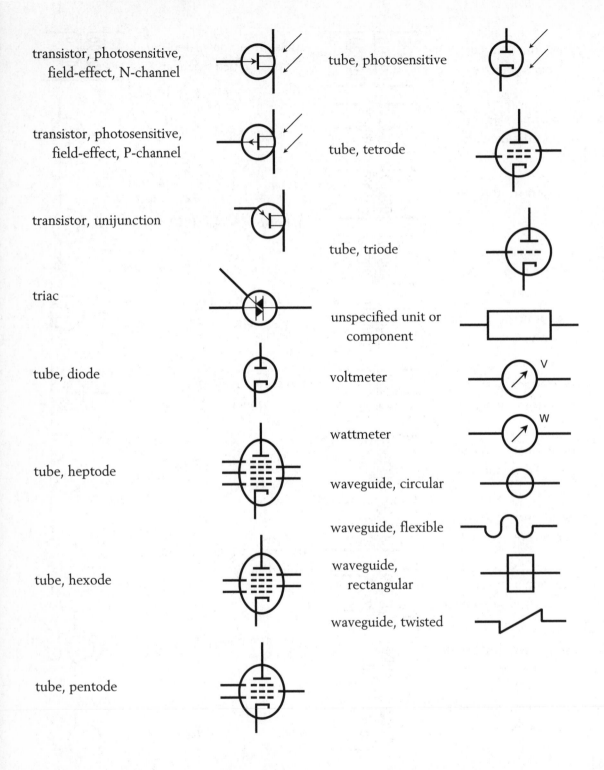

transistor, photosensitive, field-effect, N-channel

transistor, photosensitive, field-effect, P-channel

transistor, unijunction

triac

tube, diode

tube, heptode

tube, hexode

tube, pentode

tube, photosensitive

tube, tetrode

tube, triode

unspecified unit or component

voltmeter

wattmeter

waveguide, circular

waveguide, flexible

waveguide, rectangular

waveguide, twisted

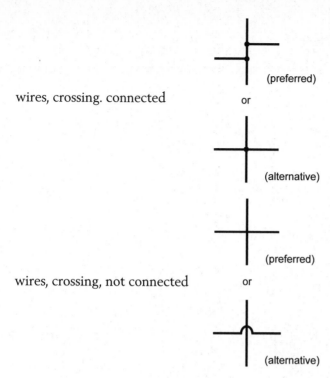

wires, crossing. connected   or

(preferred)

(alternative)

wires, crossing, not connected   or

(preferred)

(alternative)

Suggested Additional Reading

Frenzel, Louis E., Jr., *Electronics Explained*. Elsevier/Newnes, 2010.

Gibilisco, Stan, *Algebra Know-It-All*. New York, NY: McGraw-Hill, 2008.

Gibilisco, Stan, *Electricity Demystified*. New York, NY: McGraw-Hill, 2005.

Gibilisco, Stan, *Handbook of Radio and Wireless Technology*. New York, NY: McGraw-Hill, 1999.

Gibilisco, Stan and Crowhurst, Norman, *Mastering Technical Mathematics*, 3rd ed. New York, NY: McGraw-Hill, 2007.

Gibilisco, Stan, *Physics Demystified*, 2nd ed. New York, NY: McGraw-Hill, 2011.

Gibilisco, Stan, *Pre-Calculus Know-It-All*. New York, NY: McGraw-Hill, 2010.

Gibilisco, Stan, *Teach Yourself Electricity and Electronics, 5th ed*. New York, NY: McGraw-Hill, 2011.

Gibilisco, Stan, *Technical Math Demystified*. New York, NY: McGraw-Hill, 2006.

Horn, Delton, *Basic Electronics Theory with Experiments and Projects*, 4th ed. New York, NY: McGraw-Hill, 1994.

Kybett, Harry, *All New Electronics Self-Teaching Guide*, 3rd ed. Hoboken, NJ: Wiley Publishing, 2008.

Miller, Rex and Miller, Mark, *Electronics the Easy Way*, 4th ed. Hauppauge, NY: Barron's Educational Series, 2002.

Mims, Forrest M., *Getting Started in Electronics*. Niles, IL: Master Publishing, 2003.

Shamieh, Cathleen and McComb, Gordon, *Electronics for Dummies*, 2nd ed. Hoboken, NJ: Wiley Publishing, 2009.

Slone, G. Randy, *TAB Electronics Guide to Understanding Electricity and Electronics*, 2nd ed. New York, NY: McGraw-Hill, 2000.

Van Valkenburg, Mac E. and Middleton, Wendy, *Reference Data for Engineers: Radio, Electronics, Computers and Communications*, 9th ed. Elsevier/Newnes, 2001.

Veley, Victor, *The Benchtop Electronics Handbook*. New York, NY: McGraw-Hill, 1998.

Index